Local Computer Network Technologies

Notes and Reports
in
Computer Science and Applied Mathematics

Editor
Werner Rheinboldt
University of Pittsburgh

Local Computer Network Technologies

Carl Tropper

The MITRE Corporation
Bedford, Massachusetts

ACADEMIC PRESS

A Subsidiary of Harcourt Brace Jovanovich, Publishers

New York   London   Toronto   Sydney   San Francisco   1981

ACADEMIC PRESS, INC.
111 Fifth Avenue, New York, New York 10003

United Kingdom Edition published by
ACADEMIC PRESS, INC. (LONDON) LTD.
24/28 Oval Road, London NW1 7DX

Library of Congress Cataloging in Publication Data

Tropper, Carl.
    Local computer network technologies.

    (Notes and reports in computer science and applied
mathematics)
    Bibliography: p.
    Includes index.
    1. Computer networks. I. Title. II. Series.
TK5105.T76   621.3819'592   81-12671
ISBN 0-12-700850-0   AACR2

PRINTED IN THE UNITED STATES OF AMERICA

83 84   9 8 7 6 5 4 3 2

Pour Evelyne et les trois demoiselles du chateau—Lydia, Gigi, et Sabrina

Contents

Preface

The raison d'etre for this book is becoming increasingly apparent on almost a daily basis in our electronic marketplace. In particular, the interest in office automation brought forth a flood of products from such companies as Xerox, Wang, and Datapoint.

Word processing terminals, management-oriented intelligent work stations, as well as multifunction terminals incorporating both data and word processing are now being sold. All of these categories of terminals can be equipped for electronic mail, and can be tied together with a computer system to form an office automation system.

Various sources are already predicting the information age's equivalent of an industrial revolution, with the effectiveness of an office worker increasing in a manner analogous to the way in which a factory worker's productivity changed during the industrial revolution.

Office automation systems are but one example of the power inherent in the notion of a local area network. We take this term to mean a collection of processors (e.g., terminals, computers) located in the same vicinity (e.g., building) and tied together by a communications system. Other examples of local networks are medical-information networks and process-control networks.

The local networks mentioned so far involve loosely coupled processors. More closely coupled processors are the focus of an active area of research. Several companies (Honeywell, Apollo Computer) are actively engaged in the development of such distributed systems with the intention of penetrating the time-sharing marketplace (or perhaps creating a new one!).

Developments such as these are what basically lay behind the desire to write this book. The purpose of *Local Computer Network Technologies* is to synthesize the considerable amount of work accomplished in developing link access protocols for ring and bus computer-communication networks and to provide a systematic discussion of both the protocols and their associated performance models.

The basic attraction of these two approaches is their simplicity. Bus networks make use of passive taps on a cable to connect host computers or terminals to the cable via some sort of interface unit. The algorithms employed by the interface unit for gaining access to the bus* are discussed in Chapter 3. Ring networks make use of some form of buffering, however primitive, in passing messages from node to node on the network. The principle commercial approaches at the moment are the Ethernet approach championed by Xerox, DEC, and Intel, and the token passing scheme employed, for example, by Prime Computer Corporation. Ultimately both technologies are quite viable, and each will certainly find a niche in the marketplace.

In Chapter 1, the reasons behind the interest in local computer networks—technical as well as economic—are explained. An attempted definition of a local network is also provided.

Following this introductory chapter, the book divides naturally into two sections—Chapter 2 being devoted to ring networks and Chapter 3 to bus networks. Each chapter is structured along the same lines, beginning with a description of the protocols followed by a discussion of the performance models of the protocols. The performance of the protocols are then compared (to the extent that this is possible), and finally, the assumptions and weaknesses of the *models themselves* are described.

Both chapters include descriptions of many existing or planned networks, e.g., Ethernet, Mininet, Hyperchannel, and Spider. Among the protocols and networks described in the ring chapter are the Newhall (accompanied by a description of Mininet), the Pierce (accompanied by a description of Bell Labs' Spider network), Ohio State's DLCN and DDLCN, and the "Oregon State loop." The bus chapter starts off with a brief summary of the "classical" work done on the ALOHA and CSMA protocols by Kleinrock and his students, and includes a description of the fundamental delay, throughput, stability trade-off inherent to this category of bus protocols. Chapter 3 then proceeds to a description of the conflict-free schemes such as BRAM, GSMA, and DYN and closes with a description of the interesting URN protocol.

---

*Many of these algorithms allow packets to collide with one another.

Acknowledgments

I wish to thank The MITRE Corporation for its generous funding of the work necessary to expand a technical report, which I had previously written on the same subject, into this book. Special thanks go to Dave Lambert and Al McKersie for pursuing the necessary funding and to Wes Melahn for granting it.

One of the greatest debts that I have incurred in the writing of this book is owed to my secretary, Donna Leary. She patiently typed the manuscript, made a large number of corrections in the text, and single-handedly forced me to put the bibliography in order.

In this same vein, I wish to thank Jan Roussos for the job she did in proofreading the original manuscript and shepherding it through its original publication stages. I extend these thanks in spite of the fact that she forced me to write in the third person singular, and took out all of my jokes.

I wish to thank Josh Segal for explaining the MITRE cable-bus system in all its glory.

Above all, I wish to thank my family for the happiness that you have brought me (not to speak of your encouragement). I hope that this work returns some measure of this joy to you.

Chapter 1

A Perspective

INTRODUCTION

The origins of this book are to be found in a technical report written at The MITRE Corporation by the author [TROP79]. The paper was a survey of performance models of local computer networks which make use of either ring or bus technology to connect the nodes.

The report was sponsored by the U.S. Air Force, an organization for which communications plays a vital, immediate role. Hence its interest in local area networking as a new and powerful means of connecting its communications/information processing systems is rather lively.

In the commercial world, a strong interest has evolved in local area networking technology as a result of the enormous markets* which are to be developed by building the "office of the future." The object of this effort is to drastically improve the productivity of the modern business office, in an analogous manner to the way in which the productivity of the industrial factories has been increased.

The denizens of such an office would come equipped with intelligent terminals. The terminals would be interfaced via the local network with other office equipment. Facsimile devices, word processors, computers, and voice and video conferencing equipment would all be connected to the terminals.

Documents could then be produced and edited on a terminal, while intraoffice

---

*An estimation in *Fortune Magazine* places the potential market in the 6 billion dollar area!

correspondence could be carried out via "electronic memos," voice and video conferencing could be employed as part of the decision-making process, and information could be retrieved from local data bases at will. Offices in different geographical locations could be connected via a "backbone" network.

Proponents of this scenario would have us believe that as a result of these electronic advantages, decision-making would be simplified and more efficient organizational structures would emerge within the modern corporation.

Of course, the exact opposite result might occur also. Decision-making in the foregoing environment might, in fact, become extremely centralized, resulting in even more bureaucratic layers than presently exist within the modern corporation. An interesting discussion of this very point is given by Martin [MART78]. Divinity has once again seen fit to place the apple in our hands.

In any event, our office revolution would certainly appear to be underway. Word processors are now a common part of many office environments. Xerox, DEC, and Intel have recently announced a joint venture in the marketing of a local area network for use in office automation [MINI80].

Besides the very strong incentive to improve office productivity, there are two other major economic/technological factors which stand behind the emerging electronic office.

The first is the existence of inexpensive, transmission media capable of supporting a high data-transfer rate. The primary example of this media is coaxial cable. Coaxial cables can support both point-to-point or broadcast communication in the 1- to 10-/Mbit/sec range over a 1-km distance without repeaters.

The second major factor pushing us toward the electronic office is the ever cheaper, more powerful circuit designs possible by LSI (large-scale integrated) technology.

The microprocessors produced by this technology make possible the intelligent terminals employed as word processors, as well as in stand-alone fashion as part of the office complex. They also stand at the heart of the interface units used to connect the various devices to the network.

At the other end of the spectrum from the loosely coupled processors of the electronic office stands the notion of distributed processing systems. Following Stankovic and VanDam [STAN79], we take this term to mean

> A collection of processing elements which are physically and logically interconnected, with decentralized system-wide control of all resources, for the cooperative execution of application programs.

We attempted to be precise with our definition of distributed processing, as the term is too often taken in vain. A system consisting of a collection of terminals attached to a main-frame computer is not a distributed processing system.

It should be mentioned that distributed processing is very much a research area—no system corresponding to the preceding definition has been built. The *potential* advantages of distributed processing do indeed justify the excitement generated by the area. Among the advantages are the following:

1. Performance improvement, resulting largely from the inherent parallelism of distributed processing as well as load balancing. Of course, it will be necessary to be able to partition application programs in order to take advantage of these capabilities.
2. Improved fault tolerance, as failed processors could be replaced by other (functioning) processors.
3. Greater ease of system growth.
4. Improved cost/effectiveness over centralized systems due to the economics of scale inherent in the production of microprocessors.

The difficulties involved in building a truly distributed processing system are immense. One of the most vexing of these difficulties is encountered in building an operating system which must control processes having only a probabilistic knowledge of the system state. Present-day operating systems assume that global state knowledge is available to the system.

The list of open questions is, in fact, rather long. For a good discussion of the potential advantages as well as the difficulties to be encountered in distributed processing, the reader might wish to consult Stankovic and VanDam [STAN79] or Jensen [JENS78].

It should be noted in passing that several experimental systems are in the process of being built. Among these systems are the HXDP (Honeywell Experimental Distributed Processing) system, described by Jensen [JENS78]. The HXDP project, now being carried out at Carnegie–Mellon University, is focusing on the real-time control problem.

Basically, the subject of this book is the same as that of its predecessor—that is, performance models (and to some extent modeling) of local computer networks. All of our networks employ either ring or bus technology to connect the computers and/or terminals. We devote Chapter 2 to ring networks and Chapter 3 to bus networks.

With several (notable) exceptions, the modeling of local networks has *concentrated on the performance of the communications subnetwork.* The traditional measure of efficiency employed for the performance of a network has been that of time delay versus throughput.

Employing more elegant language, we are going to investigate the performance of network control protocols. These are the transport protocols employed in a computer network that are responsible for the transportation of single messages (datagrams) as well as of message streams (virtual circuits).

We are going to measure the performance of a network by the efficiency of its datagram service. We picture this performance by the time-honored delay versus throughput curve.

Before plunging ahead, it is worthwhile to reflect on the nature of performance modeling. We can do no better than to quote from Beizer* [BEIZ78] on the subject of the performance modeling of real-time software.

> The best reason for analyzing a system's performance is to exercise control over the design process. Analyses are done continually to assure that the system's performance tracks the specification. They are done to provide a rational basis for crucial trade-off decisions that may not be intuitively obvious. They are done to provide early warning of the impact of a specification or hardware change. This type of analysis is a continuing effort, mostly done by the designers themselves.

AN ATTEMPTED DEFINITION
OF A LOCAL COMPUTER NETWORK

From a functional point of view, a local computer network may be thought of as inhabiting a region between multiprocessing systems and the interconnection of geographically distributed, heterogeneous computer systems for the purpose of resource sharing. At one end of this spectrum we find an attempt to convert a collection of serial processors into parallel processors, while at the other end we discover a group of dissimilar computers tied together by a communication network, thereby enabling a user to take advantage of a variety of computing resources. Local computer networks may therefore be built to fulfill either (or conceivably both) of these design goals.

Metcalfe and Boggs [METC75, p. 395] use a taxonomy based on the parameters of bit rate and separation between computers to distinguish three types of networks:

Activity	Separation	Bit rate
Remote networks	>10 km	<0.1 Mbit/sec
Local networks	0.1–10 km	0.1–10 Mbit/sec
Multiprocessors	<0.1 km	> 10 Mbit/sec

It should be pointed out that the introduction of fiber optics technology into this area threatens to wreak havoc with the taxonomy of Metcalfe and Boggs.

---

*In the opinion of the author, this work has few parallels in the field of performance modeling of computer systems.

Another definition for local networks has been proposed by Franck [FRAN78]. He pictures a local computer network as consisting of three essential ingredients:

1. A high-speed transmission medium for data transmission over a "limited" distance. The nature of the transmission medium and the topology of the network are left unspecified.
2. Several network adapters attached to this transmission medium which serve as line interfaces for computing equipment. The adapters transmit data on the transmission medium.
3. Computing system components that can be attached to an adapter. Franck's illustration of his definition is shown in Fig. 1.

It should be noted that a panel discussion held during the third conference on local computer networks—a conference devoted to developing a definition of a local computer network—failed to achieve a definition acceptable to all participants.

Much effort has been devoted in recent years to developing technologies for local networks. Two of the basic networking technologies that have been developed are the ring and bus networks. In a ring network, the communications path is in the form of a loop. Messages originate at a source node (attached to the

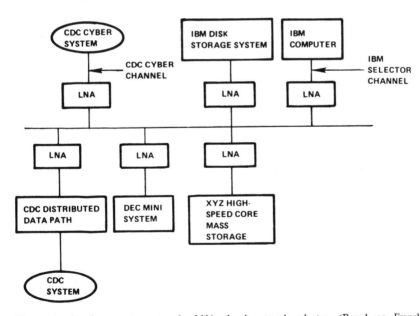

Figure 1. Local computer network; LNA, local network adapter. (Based on Franck [FRAN78].)

loop) and then flow through intermediate nodes on the way to their destination node. The intermediate nodes on the path function as relays. In a bus network, on the other hand, messages are broadcast onto a shared communications channel. Hence, all nodes attached to the communications path "hear" all the messages sent onto the loop. More detailed definitions of these networks will be presented in Chapters 2 and 3.

The reader interested in gaining an appreciation of the many issues involved in the design of a local area network would do well to read a survey of the subject by Clark *et al.* [CLAR78].

To the author's knowledge, there have been no comparisons of the performance of ring and bus networks for a given application. This is unfortunate, because recent developments in ring technology (e.g., the DLCN ring, discussed in this book) demonstrate that ring networks with impressive performance characteristics can be built.

Chapter 2

Ring Networks

INTRODUCTION

Following Anderson and Jensen [ANDE75], we characterize a ring network as a collection of processing elements (terminals or computers) that are interconnected via a communications path in the form of a loop.* This situation is illustrated in Fig. 2.

Typically, the processing element is attached to the ring by an interface device—the ring interface unit. A loop supervisor may also be present on the loop. The functions of the supervisor may include synchronization, as well as some form of flow control (to prevent the accumulation of undeliverable messages).

In general, traffic on the loop flows in one direction only, although bidirectional systems have been proposed [MAJI77, WOLF78]. Hence, each processing element receives traffic from one of its neighbors, and sends messages to its other neighbor. Messages then circulate around the loop from their source to their destination. The intermediate processing elements along the path act as relays.

Depending on the system, messages may be of fixed or variable length, and one or several messages may be permitted on the loop at a time.

---

*The words "ring" and "loop" are used interchangeably in the literature and by Anderson and Jensen.

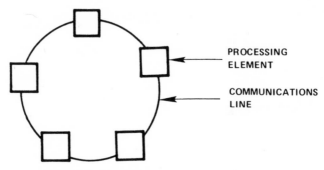

Fig. 2.   Ring network. (Based on Anderson *et al*. [ANDE75, p. 202].)

Loop networks are attractive because of their simplicity. It is fairly easy to add (or delete) processing elements without making numerous connections each time. This is a definite benefit when the network is located within the confines of an office building. In addition, start-up and system modification costs are (relatively) low.

The basic drawback of a loop system, its reliability, also stems from its simple design. An outage in either a processor or a channel can lead to disaster. Hence, it is necessary to provide some form of backup in the event of a failure. An example of such a backup would be installation of a bypass at each node, which would, in effect, delete a malfunctioning node. If the bypass is centrally activated (by a loop supervisor), it can also be used to route traffic around defective channels.

The interested reader will find a good survey of ring networks given by Penney and Baghdadi [PENN78].

LOOP ACCESS PROTOCOLS

Three basic loop access protocols have been developed: the Pierce loop, the Newhall loop, and the distributed loop computer network (DLCN). In the Pierce loop, fixed-length slots circulate around the ring. A lead field indicated to each host whether or not the next frame is occupied. In the absence of a message, a host may multiplex a message (or a portion thereof) into the available slot. Figure 3 is a diagram of this transmission mechanism.

Clearly, several messages may be sent on the network simultaneously. The principal disadvantage of this method is the fact that the messages are not all of uniform length. Some will be too short for the space allotted, resulting in a waste

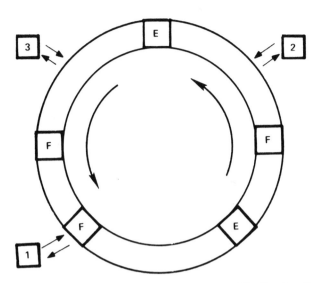

Fig. 3. Pierce loop transmission mechanism. E, empty; F, full. (Based on Reames and Liu [REAM75, p. 11].)

of space. Others will be longer than available space, necessitating software for assembly and disassembly of messages, as well as adequate buffer space.

The transmission of variable-length messages is a property exhibited by Newhall networks [FARM69]. These networks operate by "token passing." Control is passed from host to host. If a host receiving control of the loop has a message stored in its buffer, it immediately multiplexes the message onto the loop, and then passes control of the loop downstream. Clearly, simultaneous transmission of messages under these circumstances is impossible, because of message interference. Figure 4 illustrates the Newhall transmission mechanism. The basic disadvantage of a Newhall network is, again, its inability to transmit several messages on the loop simultaneously.

DLCN provides both of these advantages (the advantages of variable-length messages and simultaneous message transmission) via a store-and-forward transmission. The ring interface for DLCN consists of two buffers. The first is an output buffer which stores messages produced locally. The second is a delay buffer which buffers messages passing through the particular node in question (i.e., messages that have destinations further downstream) and inserts messages from the output buffer into the gaps between messages on the loop, as well as into the gaps produced by sinking a message at the given node. The disadvantage of this approach is that delays occur for messages as the message traverse nodes that lie on the path to their destination node.

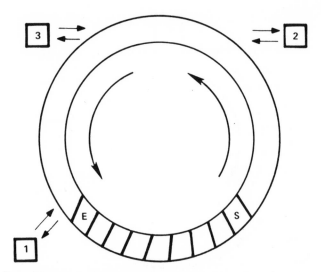

Fig. 4. Newhall loop transmission mechanism. S, start of message; E, end of message. (Based on Reames and Liu [REAM75, p. 11].)

Several architectures for loop networks have grown out of these three designs and will be described.

The Waterloo loop grew out of an attempt to prevent "loop-hogging" behavior on the Pierce network (i.e., domination of the loop by several heavy users).

The Spider network was developed with the intent of determining the feasibility of connecting a number of Pierce loops via switches to service a wide geographical area.

The DDLCN (double DLCN) is an outgrowth of the DLCN aimed at developing a fault-tolerant network. It employs two loops, carrying traffic in opposite directions, and employs the shift-register insertion technique of the DLCN on each of the loops.

The Oregon State loop, whose control is centralized, is also described. It was not influenced by the three designs just described.

MODELS

The remainder of Chapter 2 is basically devoted to a description of performance models for the ring networks that were discussed in the preceding section. An attempt was made to enliven the discussion of each of the networks by describing implementations of each of the protocols whenever possible. (Note that all of the protocols have a corresponding implementation.)

The DLCN and DDLCN are described in some detail because:

1. In the opinion of the author, the approach taken in this model—modeling the nodes as an open Jacksonian network of queues, and incorporating this with a separate model of the communication loop subnetwork—can also be employed in developing models of distributed processing systems using either bus or ring technology.

2. The DLCN is capable of supporting both variable-length messages and simultaneous message transmission.

3. Timing and management functions are completely decentralized.

It should, therefore, have a wide range of uses in various applications. For example, plans have been made for it to support a distributed data base [PARD77].

A concluding section summarizes some of the limitations and advantages of the various models, and suggests a general approach to the modeling of loop networks.

THE NEWHALL LOOP

The transmission mechanism in the Newhall loop is characterized by token passing—control of the loop is passed from host to host successively, with each host multiplexing its message onto the loop as it gains control. Farmer and Newhall [FARM69] describe the initial design and implementation of the Newhall loop at Bell Telephone Laboratories, Holmdel, New Jersey. The original system consisted of several peripherals (Calcomp plotter, teletype, etc.) attached to the loop, along with a Honeywell 516 computer employed as loop supervisor.

Since then, several local area networks have been built employing the Newhall token-passing mechanism. We shall briefly describe three of these networks.

1. Ringnet is a local area network built by the Prime Computer Corporation.

The communications subnetwork, employing the Newhall token-passing mechanism, can interface with up to 255 nodes. Shielded twinex cable is used as the transmission medium, operating at a speed of 10 MHz.

A node, pictured in Fig. 5, consists of three components—a nodal interface, a node controller, and a host computer system.

The nodal interface acts as a relay and operates under the direction of the node

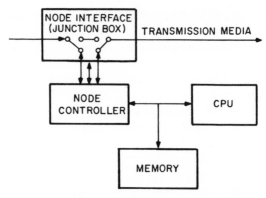

Fig. 5.   Ringnet components. (Based on Gordon *et al.* [GORD80, p. 13].)

controller. If the node controller "deselects" the interface, a straight-through connection is obtained which bypasses the node.

The Prime Computers that can act as hosts are the superminiclass 2**29 bytes of main memory, and capable of supporting up to 64 users. Examples of this class of Prime Computers are the P-400 and P-500.

Details about Ringnet are given by Gordon *et al.* [GORD80].

2. Litton has developed a local area network suitable for use in a military command and control environment.

Their distributed processing system (DPS) employs two rings for increased reliability. One of the rings is used to carry the network traffic, while the other is used solely as a backup. The technique for employing the backup loop, due to Zafiropulo, is discussed in the section on the DDLCN. In essence, the failed portion of the "live" ring is shorted out of the network and replaced by the backup loop.

The "live" loop makes use of a slightly modified version of the IBM SDLC protocol, which is, in effect, a Newhall protocol.

A four-node prototype, depicted in Fig. 6, was made operational in January 1979. For a description of the system, the reader should consult Mauriello *et al.* [MAUR79].

3. The distributed computer system (DCS) is an experimental network built at the University of California at Irvine by Farber and several associates.

The major thrust of the project was research into distributed operating systems. A system was built, employing the Newhall protocol, and connecting seven minicomputers. Although a slotted system was first built, a change was made

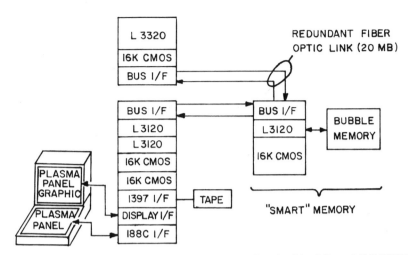

Fig. 6.   Litton distributed processing demonstration system. (Based on Mauriello *et al.* [MAUR79].)

during the course of the project to the Newhall protocol. In keeping with the objectives of the research, addressing is done by process name instead of by station address.

The interested reader may wish to consult Manning and Peebles [MANN78] for a discussion of the network and Loomis [LOOM73] for a discussion of the changeover to the Newhall protocol.

Performance Models

Following initial design, several models of the Newhall loop were developed. Two of these models determine the mean scan time of the loop, that is, the time required for the token to pass around the loop. Average waiting times are also available, but are subject to restrictions (e.g., the terminal output buffer can contain at most one message). It was not until 1977 that Carston [CARS77] developed expressions for the mean response time and mean queue lengths of a Newhall loop.

In presenting the results on the Newhall loop, therefore, we will first discuss scan-time results, based on Yuen *et al.* [YUEN72] and Carsten *et al.* [CARS77], and then the average message waiting time [KAYE72, CARS78]. We devote the following section to a discussion of the work of Labetoulle *et al.* [LABE77]. This paper is interesting because in addition to portraying the loop itself, it portrays two host processors attached to the loop, and it models both the processors and the loop as networks of queues. (A good discussion of queueing networks is given by Kleinrock [KLEI75a, KLEI76].) The present author feels that networks of queues provides a fruitful approach to the modeling of computer networks. The use of this modeling tool is particularly well illustrated in the work done on the DLCN, discussed in a subsequent section.

Yuen *et al.* [YUEN72] consider the system under consideration as a collection of (buffered) terminals, attached to a Newhall loop, which communicate with one another via fixed-length messages. Assuming Poisson input at each of the terminals, Yuen *et al.* obtain results for the mean and variance of the scan time in the case of identical (symmetric) and nonidentical (asymmetric) input traffic at each of the terminals. A critical assumption for this method is that of *light* traffic conditions. This condition is expressed mathematically in the symmetric loop case by the inequality

$$\lambda N T_s \ll 1,$$

where λ is the identical arrival rate, N the number of terminals on the loop, and T_s the service time for a terminal.

The expression obtained for the mean scan time T in the case of the symmetric loop is

$$T = \frac{N T_B}{1 - \lambda N T_s},$$

where T_B is the time delay due to the token passing (assumed to be one bit).

Other quantities of interest determined by Yuen *et al.* were formulas for the blocking probabilities at the terminals.

Simulation studies of the system were also conducted, and the results compared to the analytic results. As expected, the results were in close agreement for low traffic conditions, but mean scan time and blocking probability results diverged when the traffic became heavier.

Carsten *et al.* [CARS77] also attach a number of terminals to the loop. The terminals are represented as having infinite buffer capacity, and the loop provides service for variable-length messages. As mentioned by the authors, the infinite buffer assumption is realistic because it is inexpensive to incorporate extra memory in a terminal. In addition, if one has host computers attached to a Newhall loop, there should be ample space in the host memory to alleviate any concern about buffer overflow.

Formulas for the mean and variance of the scan time are obtained under the usual assumption of Poisson arrivals. The formula for the mean scan time $E(t_s)$ is

$$E(t_s) = \frac{D}{1 - \rho},$$

where D is the scan overhead (control character recognition, etc.), ρ the loop utilization, given by $\rho = \alpha^{-1} \sum_{i=1}^{N} \lambda_i$, λ_i the message arrival rate (messages/sec), and α the line capacity (messages/sec). Note the similarity between this formula and the usual time-delay formula for an $M/M/1$ queue.

The formula for scan-time variance is a bit more complicated. It depends on the position on the loop from which one commences the scan. Consequently, in order to simplify the calculations, approximate results which neglect the dependence for this quantity are also derived.

A four-node Newhall loop simulation produced results (mean scan time and variance) in close agreement with those obtained via the analytical model developed by Carsten et al. [CARS77].

Kaye also considers a collection of terminals attached to a Newhall loop [KAYE72]. Each terminal is assumed to have a buffer containing exactly one (fixed-length) message, resulting in the loss of messages generated while the buffer is full. (As noted earlier, this is a somewhat unrealistic assumption.) Identical Poisson arrival rates are also assumed at each terminal.

In light of these assumptions, Kaye develops an expression for the distribution of the waiting time at a terminal, defined to be the time between the loading of a message into the buffer of the terminal and the moment that its transmission commences. With this distribution in hand, expressions for the mean and variance of the waiting time are then readily obtained, as is an expression for the proportion of messages lost at a terminal during a scan. These expressions are too complex to be included herein; however, the interested reader may readily peruse them [KAYE72]. Unfortunately, Kaye conducted no simulation to verify his results.

Carsten and Posner [CARS78] develop formulas for the time delay and buffer size(s) of single and multiple Newhall loops. Closed-form expressions are developed for these quantities in the single-loop case, while approximations are developed in the multiple-loop case.

We first discuss the single-loop case, and then briefly summarize the generalization to multiple Newhall loops.

In modeling the single Newhall loops, the authors assume that N terminals are attached to the loop, with packet arrivals and message lengths governed by Poisson distributions. Each packet consists of x_i "data units," whose generating function* is given by

$$X_i(z) = E(z_i^x).$$

*The moment-generating function of a random variable x is defined as $E(e^{tx})$. We have $\left[\dfrac{d^r M_x(t)}{d_t^r}\right] = \mu_r^1$, the rth moment about the origin. The mean of x is equal to the first moment.

It is also assumed that any packets which arrive subsequent to the start of service at a terminal have to wait until the next scan for service.

Two models are developed to calculate the generating function of the queue length n_i, at each of the nodes. In describing these models, we omit the formulas for the generating functions, as they are rather complicated.

In the moving-scan model, the generating function for the n_i is the product of

1. the generating function for the total number of data units arriving during all of the overhead delays at each of the terminals during the preceding scan and

2. the generating function of the number of data units arriving during the service times at each of the terminals during the preceding scan.

In the snapshot model, the generating function for n_i is calculated by making use of the expression for n_i,

$$n_i = n_{i0} + n_{ii},$$

where n_{i0} is the number of data units in the buffer when the terminal acquires control of the line and n_{ii} the number of data units which arrive afterwards. The interested reader should consult Carsten and associates [CARS78, CARS77] for the formulas for the generating functions as well as the details of their derivations.

From a computational point of view, both of these models proved difficult to handle. Simulation results indicated that the moment-generating function of the moving-scan model could be approximated by a negative binomial form. The expression for $p_i(z)$ is

$$p_i(z) = \frac{p_i}{1 - (1 - p_i x_i(z))} \, w_i.$$

Estimates for p_i and w_i are obtained by equating the mean and variance of n_i obtained from the generating function with these estimates obtained from $p_i(z)$.

Those authors also obtain expressions for the means and variances of the following two time delays:

1. The access time t_{ai}, which is the time from arrival of a packet at a terminal to the acquisition of control of the loop by that terminal. An expression for the mean access time is given by

$$E(t_{ai}) = \frac{E(t_{si}^2)}{2E(t_s)},$$

where $E(t_s)$ is the mean scan time. $E(t_s^2)$ may be approximated via the variance (see Carsten and Posner [CARS78]).

2. The waiting time t_{wi} of a packet is the time in between the arrival of a packet and the beginning of its service. The mean waiting time is given by

$$E(t_{wi}) = (1 + \mu_i)E(t_{ai}) + d_i,$$

where $\mu_i = \lambda_i E(\lambda_i)$ and d_i is the overhead delay at the terminal, illustrated in Fig. 7. As already indicated, the authors generalized their results to the multiloop case. The system model consisted of one "major loop" with a number of minor loops attached to it via gateway terminals (see Fig. 7). A terminal on one of the minor loops communicates with a terminal on another one of the major loops by

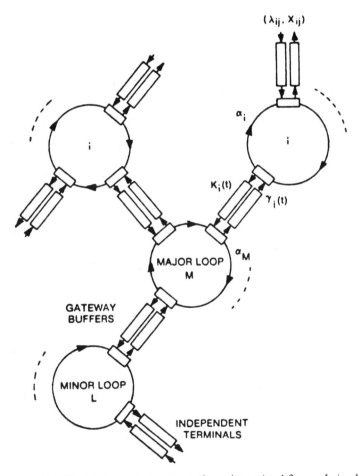

Fig. 7. Multiple Newhall loops. L minor loops of capacity α_i; $i = 1,2,\ldots,L$; $j = 1,2,\ldots,$ N_i. (Based on Carsten and Posner [CARS78, p. 44.5.4].)

first routing a message to the major loop, and then having it routed to the appropriate minor loop.

In order to develop expressions for the time delay, scan time, and queue lengths, the authors develop conditional expressions for the output process at each gateway terminal, and then apply the single-loop results already obtained.

MININET

The MININET is a two-host network of minicomputers designed to support a distributed data base. An essential feature of the data base is that it can be partitioned into components, each of which will be queried by users located in a particular geographical region. MININET was developed to support transactions processing—short queries followed by rapid responses. Credit card inquiries are an example of this sort of application.

In view of the expected bursty nature of the traffic, the designers of the system chose to implement it in the form of a two-host Newhall loop. A description of the system and of modeling work done prior to its implementation is presented by Labetoulle *et al*. [LABE77]. As the focus of the present discussion is on system modeling, we urge any readers interested in details about the hardware, operating system, etc., to consult that paper. In the process of designing the network, both analytic and simulation models of the proposed system were developed. These models are also described by Labetoulle *et al*. [LABE77].

The queueing model by Labetoulle *et al*. [LABE77] is significant. It is the first attempt (that the author is aware of) to represent the entire network (i.e., host processors in addition to the loop communication subnetwork) as a network of queues.* This formulation was employed in deriving expressions for the response time and queue lengths at the nodes. Figure 8 shows the model of the MININET network employed by the authors.

As indicated previously, there are two host processors, connected to one another via a Newhall loop. Transactions enter the system (at a rate of λ/sec) from terminals attached to the hosts, and queue for access to the host CPU (represented in the diagram by FM, file machine). The command processors, terminal processors, and message switch all reside at the file machine. As transactions arrive, the file machine determines whether or not the host CPU has the requisite data. In the event that it does, the file machine routes the request to the data host (DH).

---

*This approach pointed toward a new and potentially very effective modeling technique for ring networks which is discussed in the section on the DLCN.

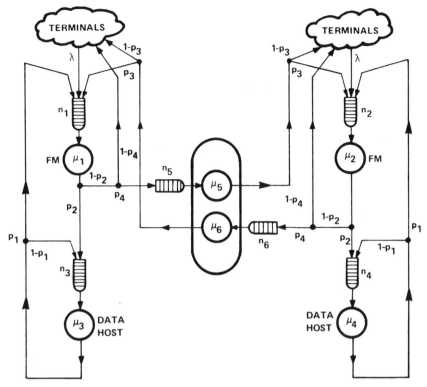

Fig. 8. Model of MININET. (Based on Labetoulle *et al.* [LABE77, p. 229].)

The data host is a separate minicomputer, in charge of secondary storage. A transaction may require several accesses, as indicated by the arrow returning to the data host. After completing the requisite number of memory accesses, post-processing is performed in the file machine, and the request is routed back to the originating terminal. In the event that a request must be satisfied remotely, it is routed onto the loop, and ultimately makes its way to the appropriate data host.

The service rates $\mu_1 - \mu_6$ are circled in the figure, while the number of items in the six queues are represented by the letters $n_1 - n_6$. The various branching probabilities are designated $p_1 - p_4$.

Figure 8 represents what is referred to in the queueing literature as an open network of queues,* a collection of nodes (service centers), with customers visiting various centers guided by a matrix of node-to-node transition probabilities.

---

*An excellent summary of networks of queues and their applicability to the modeling of computer–communications networks is presented by Kleinrock [KLEI75a, KLEI76]. The interested reader is urged to consult these volumes for a survey of the state of the art in this area.

The network is called ''open'' because customers are permitted to enter and exit from the system at the individual nodes.

Hence, the arrival rate at the ith node λ_i may be written as

$$\lambda_i = \gamma_i + \sum_{j=1}^{N} r_{ji}\gamma_j,$$

where γ_i is the external arrival rate at node i, λ_j the arrival rate at node j, and r_{ij} the transition probabilities.

Jackson established that for such a network, one might obtain the distribution for the number of customers in the system by multiplying the individual distributions at the nodes [JACK63]. This assumes a Poisson arrival rate at the nodes as well as exponential service times. Hence, if $p(k_1, \ldots, k_n)$ represents the probability of k_1 customers at node 1, k_2 customers at node 2, etc., an expression for $p(k_1, \ldots, k_n)$ (in our case) is provided by

$$p(k_1, \ldots, k_n) = \prod_{i=1}^{N} (1 - p_i)p_i^{k_i},$$

where $p_i = \lambda_i/\mu_i$, the utilization of the server.

More general results are presented by Muntz and Basket [MUNT72] and Basket et al. [BASK75]. Labetoulle et al. [LABE77] used these more general results (in particular, the concept of local balance) in deriving the foregoing distribution, perhaps unnecessarily.

Labetoulle et al. [LABE77] compute the arrival rates λ_i by solving the following equation, which corresponds to the equation already presented for arrival rates:

$$
\begin{bmatrix} \lambda_1 \\ \lambda_2 \\ \lambda_3 \\ \lambda_4 \\ \lambda_5 \\ \lambda_6 \end{bmatrix}
=
\begin{bmatrix} \lambda \\ \lambda \\ 0 \\ 0 \\ 0 \\ 0 \end{bmatrix}
+
\begin{bmatrix}
0 & 0 & p_1 & 0 & 0 & p_3 \\
0 & 0 & 0 & p_1 & p_3 & 0 \\
p_2 & 0 & 1 - p_1 & 0 & 0 & 0 \\
0 & 0 & 0 & 1 - p_1 & 0 & 0 \\
(1 - p_2)p_4 & 0 & 0 & 0 & 0 & 0 \\
0 & (1 - p_2)p_4 & 0 & 0 & 0 & 0
\end{bmatrix}
\begin{bmatrix} \lambda_1 \\ \lambda_2 \\ \lambda_3 \\ \lambda_4 \\ \lambda_5 \\ \lambda_6 \end{bmatrix}.
$$

All of the preceding parameters can be obtained by assumption or experimentation, except for the loop service parameters μ_5 and μ_6. (The estimated service rate of the file machine provides us with values for μ_1 and μ_2, for instance.) To obtain these parameters, the authors model the Newhall loop as follows:

Let μ_L be the line service rate. If $n_6 = 0$, that is, the corresponding port is idle, then $\mu_5 = \mu_L$. If $n_6 \neq 0$, that is, this port is not idle, then $\mu_5 = \mu_L/2$.

Hence, if q is the probability that the port with service rate μ_6 is idle, then

$$\mu_5 = q\mu_L + (1 - q)\frac{\mu_L}{2} = \frac{\mu_L}{2}(1 + q).$$

To express q in terms of μ_L, the authors assume that the probability that the loop is idle (q^2) can be expressed as

$$q^2 = 1 - \frac{2\lambda}{\mu_L}.$$

This corresponds to approximating the loop by an $M/M/1$ queue with a total arrival rate of 2λ. (In such a queue, the probability of the server being idle is $1 - \rho$, where ρ is the utilization of the server.)

Substituting this expression for q in the previous equation, we obtain

$$\mu_5 = \frac{\mu_L}{2}\left[1 + \left(1 - \frac{2\lambda}{\mu_L}\right)^{1/2}\right].$$

Expressions for response time and queue lengths at the various service centers can now be obtained by noting that in this (Jackson's) model, each service center behaves as an independent $M/M/1$ queue.

A simulation model of the network was also created, and its results were compared with the queueing model. Response time and queue lengths at various servers were the primary quantities of interest in this comparison. The program is an event-stepped simulation written in SIMSCRIPT. The model on which it is based is the queueing network displayed earlier, combined with—as the authors put it—a number of ''refinements'' to reflect the simplifications inherent in the queueing model (e.g., message transfer protocols).

For two-host networks, the results of the queueing and simulation models are in close agreement. Agreement is especially good if the fraction of traffic at any node which goes remote is assumed to be less than 0.6. (This is within the operating range of the network.)

The queueing model was generalized to more than two hosts and the results were compared to the corresponding simulation output. In this case, serious divergence occurred when the fraction of remote traffic exceeded 0.4. This is to be expected, as the simple two-host model created by Labetoulle et al. [LABE77] does not admit to a straightforward realistic generalization. The literature does not contain a good model of the Newhall loop with more than two ports.

THE PIERCE LOOP AND ITS RELATIVES

The transmission mechanism of the Pierce loop consists of multiplexing a message into one or more fixed-length time slots which circulate continuously around the ring. The loop was first proposed by Pierce [PIER72a] to accommodate a population of users that generate traffic characterized by a high peak-to-average ratio, that is, ''bursty'' traffic. Inquiry–response systems, such as credit card verification and electronic funds transfer, are examples of systems expected to support a bursty population. Pierce has also suggested the possibility of nationwide loop networks which would consist of a large national loop attached to several regional loops, attached in turn to local loops to be used as the access mechanism to the network [PIER72b].

The performance of the Pierce loop has been discussed by Hayes and Sherman [HAYE71] and Anderson *et al.* [ANDE72]. Hayes and Sherman developed two analytical models of the system and compared their predictions to those produced by a GPSS simulation. Details of this simulation and the results of the studies conducted with it are given by Anderson *et al.* [ANDE72].

The two models developed by Hayes and Sherman [HAYE71] were intended to complement each other. The first model portrays a population of bursty users, while the second focuses on sources that tend to generate longer messages. Both models depict a collection of data sources attached to the ring. Each message is then divided into a collection of fixed-length packets, and awaits its turn to be multiplexed onto the loop. The output of each of these data sources consists of alternating active and idle periods. It is assumed that the lengths of the active and idle periods are exponentially distributed and statistically independent of one another. Both models first develop an expression for the length of the busy and idle periods on the line, and then develop an expression for the average time delay based on these calculations.

As already mentioned, the first model portrays a collection of bursty users on the ring. By assumption, the lengths of the busy and idle periods at each individual source are exponentially distributed with known mean. It may be shown that the resultant idle period on the line (as seen by an arbitrary station on the ring) is also exponentially distributed. Its mean is equal to the sum of the mean values of all the ''source'' idle periods, that is, of all the traffic generated by individual nodes feeding traffic into the chosen node. By using this method, the lengths of both the busy and idle periods can be calculated. To calculate time delay, Hayes and Sherman rely on a queueing model that views the line as a server subject to periodic breakdown [AVI63]. Line busy periods are interpreted as periods during which the server is active; idle periods are interpreted as breakdowns. The interested reader should consult Hayes and Sherman [HAYE71] for details of the calculations as well as for the expressions for time delay and line idle and busy periods.

The second model developed by Hayes and Sherman [HAYE71] is oriented toward more active sources on the line. As such it assumes that the mean length of the line idle period at the input to a particular station on the ring is known. A further assumption is that the data flow from a station is at a constant rate equal to the average rate. An expression is developed for computing the length of the idle and busy periods at the output of the station, and this expression is applied in iterative fashion around the ring. Based on work by Sherman, the probability density function of the contents of the buffer at an arbitrary station is derived [SHER70]. An application of Little's theorem [KLEI75c] yields an expression for the time delay.

Hayes and Sherman also developed a simulation model of the ring, which was run with 10, 50, and 100 stations on the ring at various line loadings. The simulation model assumed a symmetric traffic pattern (intended to represent the traffic situation that might be encountered in a national ring). The results obtained with the analytical models compared well with the simulation results (especially for moderate line loadings, below 0.5), except that the analytical models produced more conservative time-delay values.

Spider

Spider is an experimental data communications network which was built at the Bell Telephone Laboratories (Murray Hill, New Jersey) under the direction of A. G. Fraser. A detailed description of the network is given by Fraser [FRAS74].

This network was built with the notion of investigating Pierce's idea of interconnecting a collection of (Pierce) loops via switches [PIER72a]. The switches were to be employed to manage the message flow between the loops. It will be recalled that Pierce [PIER72a] proposed a national data communications network which consisted of a collection of Pierce loops arranged in a hierarchical fashion. A national loop was connected to a number of regional loops which were in turn connected to local loops.

Although Fraser initially planned to build several loops, only one (along with a switch) was actually constructed. It connects 11 minicomputers of five different types.

Figure 9 is a diagram of the loop. The LAMs (line-access modules) are used to isolate failed terminals from the loop.

Each terminal or computer is connected to the loop via a terminal interface unit (TIU). The loop is implemented as a Bell System T1 standard, operating at 1.544 Mbit/sec. The TIU, which contains a microcomputer by the name of Fly, has as its major function the control of flow between the line and the attached terminal (or computer). The switch is implemented as a TEMPO I minicomputer.

If a terminal wishes to send a message to another terminal, it first sends a signal packet to the switch indicating its intention. The switch, in turn, establishes a virtual channel by notifying the destination terminal that it should expect a message, and then records the IDs of the correspondents in a table for all future transactions. Should the destination terminal be busy, packets from the source terminal

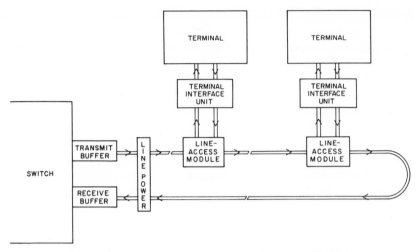

Fig. 9. Functional model of Spider. (Based on Fraser [FRAS74].)

are stored in the switch (if there is room). In the event that the destination terminal has several source terminals wishing to communicate with it, it is free to choose the order in which to accept their transmissions.

The purpose of choosing a virtual circuit mode of communication was to minimize the addressing information required in a packet. This minimization becomes especially valuable in the event that there are multiple loops.

In the course of developing Spider, a number of questions were raised as to the choice of the multiplexing strategy to be employed as well as to the storage requirements in the central switch. To answer these questions, both mathematical as well as simulation models were developed. A description of these models forms the content of the next section.

Performance Models

We start with a discussion of the analytical and simulation models employed to compare the line multiplexing strategies, and then present a summary of the simulation models used to determine buffer requirements.

Three multiplexing strategies were studied:

1. adjacent-slot seizure demand multiplexing,
2. alternate-slot seizure demand multiplexing, and
3. synchronous time-division multiplexing (STDM).

In demand multiplexing (DM), the user multiplexes a packet into a free time slot. If the user is permitted to make use of contiguous time slots, it is referred to as being adjacent-slot seizure DM. If the user is permitted access to alternate slots, the technique is referred to as alternate-slot seizure DM.

In STDM, each user is assigned a fixed allocation of time slots that recur periodically.

Queueing models were developed for STDM and DM with adjacent-slot seizure. As alternate-slot seizure proved to be intractable; it was simulated. Figure 10 represents the model employed in all of the studies.

A Poisson distribution governs the arrival and departure of messages at the terminals. The distribution has a mean of messages per second. Return message flow from the switch is also assumed to be governed by a Poisson distribution with a mean of messages per second (i.e., the switch does not affect message traffic).

The time delay is defined by Hayes as being the time elapsed between the arrival of a message at a terminal and the departure of the last packet of the message on the line.

We first describe the queueing models for STDM and adjacent-slot DM, and thus follow this description with a description of the simulation models.

In STDM, each user is assigned one slot per "cycle," which is defined to be $T_c = NT_s$ sec in length. The duration of a slot is T_s. If a message containing M_{L+1} packets arrives w sec before the end of the slot, then it must wait $w + (m_{L+1} - 1)T_c$ sec before being transmitted. If L packets arrived during the time interval $T_c - w$, and if q_t packets were held over from preceding cycles, then the expression for time delay is

$$D = qT_c + T_c \sum_{i=1}^{L} m_i + w + (m_{L+1} - 1)T_c,$$

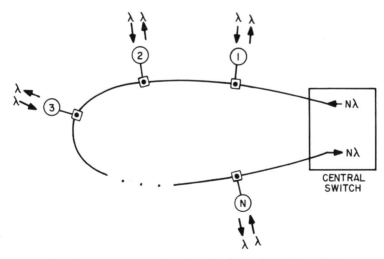

Fig. 10. Loop configuration. (Based on Hayes [HAYE74, p. 221].)

where m_i is the number of packets in the ith message. Hayes develops the moment-generating function for the preceding formula, from which he obtains the mean time delay

$$\bar{D} = T_c(\bar{m} - \tfrac{1}{2}) + \frac{T_c^2 \overline{m^2}}{2(1 - T_c \bar{m})} ,$$

where $\bar{m}$ is the mean message length in packets.

In the case of DM with adjacent-slot seizure, Hayes basically employed the same model which he used to analyze the Pierce loop. Message flow past each terminal is looked upon as consisting of alternate idle and active periods. These periods are looked upon as being caused by the Poisson arrival of messages at the switch, an assumption which is justified under light loads. The busy period at the output of the switch is modeled as the busy period of an $M/G/1$ queue, while the message interarrival time is governed by a Poisson distribution. Since $(N - 1)\lambda$ messages/sec pass by each terminal, the mean idle period is $1/(N - 1)\lambda$ sec.

By making use of Avi-Itzhak's model of a server subject to periodic breakdowns [AVI63], an expression for the message delay may be found.

Since calculation of the line busy and idle periods for alternate-slot seizure is more difficult than for adjacent-slot seizure, it was necessary to simulate this case.

Relying on assumptions already described, Hayes developed a set of curves depicting the delay versus throughput performance of both the adjacent as well as alternate-slot seizure strategies. The curves are pictured in Fig. 11. As can be seen, the analytical and simulation results are in close agreement for adjacent-slot seizure.

The principal result indicated by Fig. 11 is the somewhat better performance of the adjacent-slot seizure technique. Measurements later confirmed these results. As the difference in performance between the two techniques is not great, Hayes suggests that a choice between the two strategies should probably rest on their relative ease of implementation.

The comparison between DM and STDM (Fig. 12) yielded the result that DM is quite superior to STDM for all line loadings. As expected, the superiority is most pronounced at low loadings.

It will be recalled that the central switch stores packets that cannot be immediately delivered to their destinations. In the event that the central switches storage is used up, source terminals are prohibited from further transmissions. In an attempt to forestall any throughput problems, the buffering strategy to be employed (i.e., dedicated or common) was studied via simulation.

In dedicated buffering, it was assumed that there were fixed-size buffers, each accommodating a separate terminal. When a buffer was filled, the transmitting terminal was inhibited from sending any further packets. In the case of common buffering, a first-come first-serve service strategy was pursued (thus allowing one terminal to monopolize the buffer).

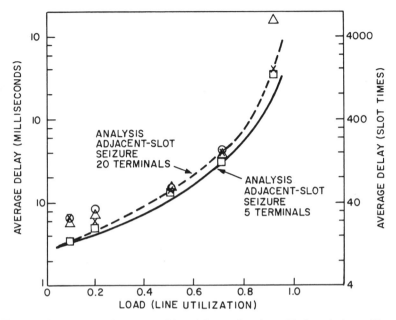

Fig. 11. Simulation results, average delay. Adjacent-slot seizure: □, 5 terminals; ×, 20 terminals. Alternate-slot seizure: △, 5 terminals; ○, 20 terminals. (Based on Hayes [HAYE74, p. 239].)

Results of the simulation indicate that the difference in throughput between the strategies is marginal except when small amounts of buffering are used.

A simulation study was also done to see if the strategy employed by the receiving terminal in switching between buffers would affect the amount of storage in the central switch. No significant difference was discovered. For more details on these studies, the reader is advised to consult Hayes [HAYE74].

Waterloo Loop

Yu and Majithia* [YU79] propose a "prioritized" Pierce loop in which the priority of a node for access to the loop is determined by the queue length of the packets waiting to be transmitted at the node. Their reasons for proposing this loop rest on oft-repeated criticisms of the Pierce loop:

1. A disproportionate amount of bandwidth is wasted in acknowledging packets. A full slot is required for an acknowledgment. This same point applies to other control messages.

*Both authors are associated with the University of Waterloo in Ontario, Canada—hence the name of the loop.

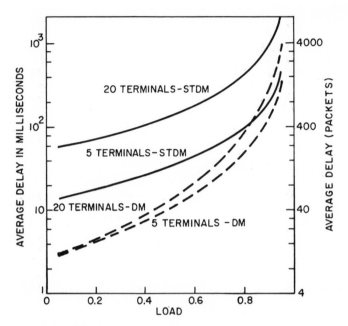

Fig. 12. Average delay versus loading in DM and STDM (30% of the messages are 32 packets long). (Based on Hayes [HAYE74, p. 241].)

2. Several heavy users can effectively dominate the loop, depriving their more "bursty" brethren of access to the loop.

3. A strong positional influence is present; that is, users closer to the loop supervisor have first access rights to the system.

In order to better equalize access to the net, Yu and Majithia define a prioritized access mode to the loop. They suggest a full-duplex Pierce loop with (alternating) control and data slots. The slot sequence on the loop is illustrated in Fig. 13. Each node maintains independent queues for control and data packets. Access rights to the loop are determined by the queue length of the data packets by designating certain queue lengths to be transition points to a higher access priority. For example, a four-priority system would assign its priorities as follows:

The priority of the slot is   1   2   3   4
if the queue length >   0 N_1 N_2 N_3 packets

Slots are also assigned priorities for use by the appropriate node(s). In their protocol, Yu and Majithia allow a slot of priority i to use slots of priorities $1, \ldots, i - 1$. When one of the preceding thresholds is passed at a particular node,

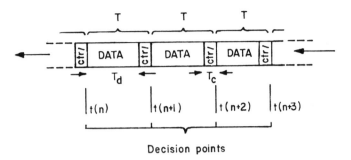

Decision points

Fig. 13. Slot sequence on the loop.

the node sends a control packet to the loop supervisor, signifying a change in its priority. The loop supervisor then generates the number of slots for each of the priority levels proportionate to the number of potential "clients" for that priority slot. If, for example, x_1, x_2, x_3, x_4 are the number of nodes of priority 1–4, then the number of nodes which can use a slot of priority 1 is $Y_1 = x_1 + x_2 + x_3 + x_4$. The number of nodes which can use a slot of priority 2 is $Y_2 = x_2 + x_3 + x_4$, etc. Hence the ratio in which the slots are generated is $Y_1 : Y_2 : Y_3 : Y_4$. The ratio can also be a biased one, for example, $Y_1 : aY_2 : bY_2 : cY_2$, although Yu and Majithia do not consider this case.

While it is true that this protocol is aimed at giving the heavy user his due, the *system is not completely decentralized*. The loop supervisor is absolutely necessary for the access protocol to function.

Some of the impetus for the development of a prioritized access protocol rests on previous work done in this direction by Katz and Konheim [KATZ74] and by Zafiropulo [ZAFI73].

Performance Models

Both queueing and simulation models were developed for the purpose of analyzing and comparing the performance of the loop to that of the (no-priority, half-duplex) Pierce loop and to the Newhall loop.

The queueing model addresses the delay versus throughput performance of the loop. The simulation studies were conducted in order to confirm and extend the queueing results as well as to perform the comparison between the proposed loop and the Pierce and Newhall loops. We present a brief summary of both the queueing and simulation studies.

Queueing Model

Yu [YU79] analyzes only the data traffic, referring the reader to [YU76] for details of the control-packet analysis. A 16-node network is considered, with the traffic load uniformly generated and with the arrival rate governed by a Poisson distribution. Messages lengths are exponential (mean length = 1000 bits) and packet lengths are 650 bits. Channel capacity is 50 kbit/sec.

The object of the queueing analysis was to obtain a (recursive) expression for the queue lengths at fixed points in time (Fig. 13). Once obtained, expressions for mean queue length and mean packet delay can be readily derived from this expression.

A fundamental problem in doing the queueing analysis is the fact that the access priority for the line alternates from node to node, depending on the queue lengths at the nodes. Because the problem of alternating priority queueing disciplines is not a tractable one, Yu and Majithia were forced to resort to bounding the performance of their loop by three limiting cases:

1. *No relative priorities* All nodes remain at the same priority; essentially a full-duplex Pierce loop.
2. *Best case* All nodes are kept, irrespective of their load, at the lowest priority and only a fixed node can increase its priority.
3. *Worst case* All nodes are kept at the highest priority irrespective of the load. The fixed node is the only one that can change its priority.

The behavior of the loop lies in between the behavior of the last two cases.

The same problem of dealing with alternating priorities in the queueing structures surfaces in the modeling of the DLCN. It is circumvented by approximation.

The analysis starts by first considering a half-duplex loop, and then proceeds to the full-duplex case. We follow the author's presentation.

Figure 13 represents the sequence of slots containing control and data packets that are confronted by a node. Decisions by a node as to whether it can use a slot or not are made at the leading edge of the slot, and are referred to (appropriately enough) as decision points. They are represented in Fig. 13 by $t(n)$, $t(n + 1)$, etc. Making use of this diagram, Yu and Majithia developed the state equation (for an explanation of the symbols, refer to Table 1)

$$p(k, n + 1) = P(k + 1, n)*\alpha(0)*p(k + 1)$$
$$+ P(K,n)*(\alpha(0)*q(k) + \alpha(1)*p(k + 1))$$
$$+ P(k - 1,n)*(\alpha(1)*q(k) + \alpha(2)*p(k + 1))$$
$$+ \cdots + P(1,n)*(\alpha(k - 1)*q(k) + \alpha(k)*p(k + 1))$$
$$+ P(0,n)*(\alpha(k)*q(k) + \alpha(k + 1)*p(k + 1)).$$

The boundary condition at $k = 0$ is given by

$$P(0,n + 1) = P(0,n)*(\alpha(0) + \alpha(1)*p(1))$$
$$+ P(1,n)*\alpha(0)*p(1).$$

TABLE 1

Symbol Definitions

$P(k,n)$	Probability that at the nth decision point the queue length is k; under steady state, the probability is independent on n such that $P(k,n)$ is represented by $P(k)$
$P(k)$	Probability that an arriving slot is available for use when the data queue length just prior to the decision point is k
$\alpha(i)$	Probability that i packets arrive into a data queue between two decision points
$q(k)$	Probability that an arriving slot is *not* available for use when the data queue length just prior to the decision point is $k = 1 - p(k)$

The objective at this stage is to develop expressions for $\alpha(i)$ and $p(i)$ in the preceding equation. It is, of course, possible to solve the recursive relationship for the limiting cases.

An expression for the probability $\alpha(i)$ that n packets arrive at a node between two decision points which are τ time units apart is given by

$$\alpha(i) = \text{Prob}(n \text{ packets arrive in } \tau)$$
$$= \text{Prob}(1 \text{ message arrives and contains } n \text{ packets})$$
$$+ \text{Prob}(2 \text{ messages arrive and total } n \text{ packets})$$
$$+ \cdots + \text{Prob}(n \text{ messages arrive and total } n \text{ packets}).$$

Making use of the exponential message length and message interarrival time assumptions, it is a relatively easy matter to obtain expressions for each of these probabilities as well as for their z transforms. Once done, one has only to differentiate the sum of z transforms and to evaluate the result at $z = 1$ to obtain the mean number of packets arriving at a node between two decision points. Details of the derivation are given by Yu and Majithia [YU79].

In order to derive an expression for $p(k)$ (the probability that an arriving slot is available for use given a queue length of k prior to the decision point), the authors assume a uniform traffic distribution between nodes, and statistical equilibrium.

For the no-relative priority case, the authors develop an expression of the form

$$p = 1 - \sum_{j \neq i} \lambda_t(i, j)\tau,$$

where $\lambda_t(i,j)$ is the traffic passing through station i (the observed station) contributed by station j. They also show that $\lambda_t(i) = \frac{1}{2}\lambda M(\frac{1}{2}N - 1)$.

This expression must be generalized to include the priority of the nodes relative to the rest of the network. Define $f(j: S, k)$ as the probability that a slot arriving at node j may be used by j given that the network is in state S and the queue length at j is k. Then the probability $p(k)$ that station i may use an arriving slot given that its queue length k is given by

$$p(k) = \left[1 - \sum_{j:j \neq i} \lambda_t(i,j)\tau f(j: S, k) \right] f(i: S, k).$$

By evaluating $f(i: S, k)$, expressions for $p(k)$ corresponding to the remaining two priority cases are developed by the authors.

Armed with expressions for $\alpha(i)$ and $p(k)$, the next step is a generalization to the full-duplex loop. Here the authors assume that the decision points are uniformly separated in time, with a mean separation of $\tau/2$. They then solve the recursive relationship for $p(i)$ for each of the three boundary conditions.

In the no-relative priority case, the expression for the mean queue length agrees with that obtained by Hayes and Sherman [HAYE71].

Figure 14 portrays the behavior of the mean queue length versus the network traffic intensity, defined to be $N\lambda\tau M$, where N is the number of nodes in the network and M the number of packets in a message. The B, W, and NRP curves represent best, worst, and no-relative priority curves with the transition points set at 5, 15, and 25. The performance of the loop is, as previously indicated, bounded by the B and W curves. Note that the condition for stability is that $N\lambda\tau M < 4$.

The authors indicate that the foregoing analytical results are confirmed by simulation studies, the details of which are presented by Yu [YU76].

Simulation studies indicate that the network is indeed effective in handling unbalanced traffic (e.g., inquiry–response-type traffic, in which a single host dominates the traffic statistics). In Fig. 15 the authors retain the same transition points, but plot the mean queue length at node i. They assume that all nodes on the curve except i are kept at priority 1. The W curve assumes other nodes are kept at priority 2. In the figure, λ_i is the arrival rate at node i, while λ_j is the (fixed) arrival rate at the other nodes.

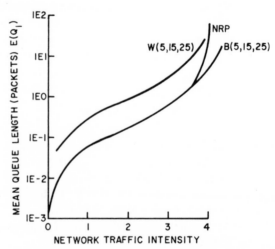

Fig. 14. Mean queue length at a station under uniform traffic condition. (Based on Yu and Majithia [YU79, p. 101].)

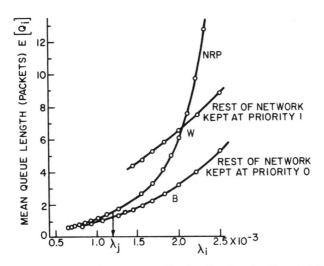

Fig. 15. Priority structure performance under biased traffic. (Based on Yu and Majithia [YU79, p. 101].)

As is clearly indicated in Fig. 15, the performance of the proposed loop surpasses the no-relative priority (Pierce) loop.

Comparison with Pierce and Newhall Loops

Simulation studies were conducted to compare the performance of the proposed loop to that of the Pierce and Newhall loops. The original Pierce and Newhall protocols were employed on a 16-node full-duplex loop having a total capacity of 50 kbit/sec. The proposed loop was simulated using the "no-relative priority" protocol, that is, it was essentially a full-duplex Pierce loop.

Figure 16 indicates that the delay for the proposed loop lies between the delays for the Pierce and Newhall loops at low to medium traffic levels. The significant point about the loop is that it saturates at much higher traffic levels than either the Pierce or Newhall loops.

The Newhall loop has a performance superior to that of the Pierce loop. The reason for this is its ability to carry variable length messages. In the Pierce loop, messages must be "packetized" prior to their transmission, thus increasing their delay. Message acknowledgments also consume a large amount of bandwidth on the Pierce loop.

The average buffer size and control packet delay of the proposed loop were also established as being smaller than either the Pierce or Newhall loops. Simulation results also portray the performance of the proposed loop as being superior to that of its competitors under biased traffic conditions.

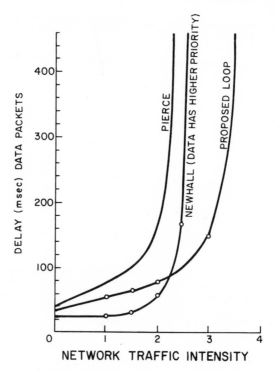

Fig. 16. Data packet delay versus network traffic intensity for various loop structures.

For more details on the analytic or simulation studies, the reader may wish to consult Yu [YU76].

DLCN

The distributed loop computer network (DLCN) was designed and is presently being implemented at Ohio State University.* The design of this network is documented in a series of papers by Liu, Babic, Pardo, and Reames. Our description of the network and the associated queueing models is based on the work of Reames and Liu [REAM75] and Babic *et al.* [BABI77]. The DLCN has been designed to provide

---

*In a telephone conversation with the author, Professor Liu, who is associated with the project, indicated that he expects to have a prototype network (three nodes) completed by the end of summer 1981.

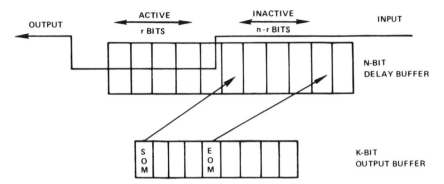

Fig. 17. Model of DLCN ring interface. (Based on Reames and Liu [REAM75, p. 11].)

1. simultaneous message transmission between hosts and
2. transmission of variable length messages.

The network thus combines the advantages of the Newhall network (item 2) and the Pierce network (item 1). The DLCN provides both of these advantages via a store-and-forward solution. The ring interface employed to effect this solution consists of two buffers, an output buffer that is used to store messages produced locally and a delay buffer. The delay buffer stores messages passing through the particular node in question (i.e., messages that have destinations further downstream) and inserts messages from the output buffer into the gaps in between messages on the loop, as well as the gaps produced by sinking a message at the given node. A diagram of these two buffers (Fig. 17) will help explain their operation.

We first consider the operation of the delay buffer. Bits arrive serially at the buffer along the incoming line, one per time unit. Assume that an on-going r-bit message arrives at times $t_0, t_1, \ldots, t_{r-1}$. These bits will be stored in the first r positions of the buffer, the active portion. As the last bit of this on-going message arrives, the first bit will be transferred onto the output line. In the event that another on-going message does not appear for another time unit, the active portion of the delay buffer is reduced by one bit. (Our first-in, first-out queue is reduced by one customer.) As the gaps between messages continue to appear, the size of the delay decreases—approaching its irreducible minimum, which is one.

In the event a message of length s bits has been assembled in the output buffer and a gap appears between messages on the input line, the s-bit message is parallel-transferred into the inactive portion of the delay buffer immediately adjacent to the active portion. This transfer occurs under the provision *that there are at least $s + 1$ bits available in the inactive portion of the delay buffer.* The active portion of the delay buffer now contains $r + s + 1$ bits, while the remainder of the delay buffer now constitutes the inactive portion. The extra bit is used for delaying new, incoming messages. This is the conceptual mechanism employed for inserting messages into the gaps between messages on the input line.

In the event that the length of the message in the output buffer exceeds the space available in the delay buffer, it does not gain access to the delay buffer until the active portion is sufficiently reduced. This tactic clearly penalizes a heavy user of the system. However, the length of the delay buffer can be a design variable, capable of being increased to favor certain important users. The basic design tradeoff brought to light is that of balancing the ability to output messages (achieved by according the size of a user's delay buffer to his importance).

A more detailed description of the operation of the ring interface transmitter, as well as several possible hardware implementations of the transmitter, is presented by Reames and Liu [REAM75].

Modeling

Before the DLCN was implemented, both analytical and simulation studies of the network were conducted. We intend to focus on the queueing model because it provides a framework for the modeling of other local networks. Comparisons of the analytic and simulation results will be included in the discussion. We therefore devote this section to a detailed summary of the model, based on the work of Babic *et al.* [BABI77] and Liu *et al.* [LIU77].

Two fundamental points about the DLCN queueing model are the following:

1. It portrays each node as an (open) Jacksonian network of queues. (A description of queueing networks is given by Kleinrock [KLEI76, pp. 212-236].)
2. It approximates the loop communications subnetwork as a single-server queue.

Thus, the average time delay for the entire network can be calculated by computing the time delays at individual (host) nodes via a Jacksonian model and determining the loop* delay based on the model described by Liu *et al.* [LIU77].

For several reasons, the network was not modeled by simply appending queueing submodels for the hosts onto the model for the loop subnetwork. The most important reason, in the opinion of the author, is that this approach would have resulted in an unwieldy number of equations. By employing an expression for the average subnetwork time delay, the authors greatly reduced the complexity of the model.

Two other difficulties that appear in the modeling of the loop subnetwork,

---

*We risk some semantic confusion by writing ''loop'' for ''loop communications subnetwork'' in what follows.

which also mitigated against this approach, will be discussed in the next section. Briefly, they are as follows:

1. The loop service discipline alternates access priority to the loop between incoming messages and locally generated messages.
2. A message is simultaneously serviced by more than one node.

Ultimately, the authors derive expressions for

(a) channel and processor utilizations,
(b) queue lengths at the processors and at the channel, and
(c) mean time delay in the network.

In describing their work, we first summarize the model for the entire network [BABI77], and then we summarize the model for determining the loop time delay formula [LIU77].

Conceptual Models

Figures 18–20 depict the conceptual basis for modeling the DLCN. It is assumed that interarrival times and service times are distributed exponentially and that service follows a first-come, first-serve (FCFS) queueing discipline.

Figure 18 shows a simplified model of the host and its relationship to the network. As the figure shows, terminals that generate requests and receive responses at rate λ_i are attached to the ith host. The message streams A_i, B_i, and C_i, by which the host interacts with the rest of the network, are also indicated.

Focusing on the host, we note that it is conceptually divided into a *communi-*

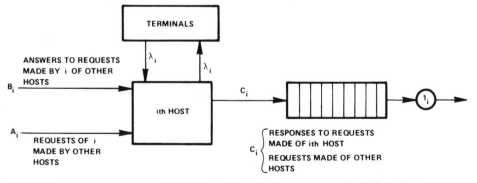

Fig. 18.   Model of host message flows. (Based on Babic *et al.* [BABI77, p. 74].)

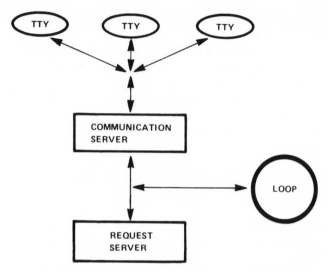

Fig. 19. Model of ith host. (Based on Babic *et al.* [BABI77, p. 74].)

cations server and a *request server* (Fig. 19). The communications server receives requests from terminals, performs preprocessing functions on the requests, and either routes them to the local host if they can be satisfied locally or routes them to a remote host if they cannot. The communications server also receives both local and remote responses to requests, performs postprocessing on them, and returns them to the terminals that made the inquiries.

Figure 20 shows a detailed flow of messages through the host and into the network. In this figure, requests are seen entering the communications processor (at rate λ_i) from the terminals attached to the ith host. Two other streams also seen entering the communications processor are

1. stream B_i, which consists of responses to (remote) requests made by the ith host, and
2. a return stream of responses from the request server of the ith host.

Both of these streams will undergo postprocessing and will then be routed to the appropriate terminals. The communications processor is depicted as server 2, and processes requests at rate μ_{2i}. If the requests can be satisfied locally, they are passed onto the request server, represented as server 3. The request server processes this stream at rate μ_{3i}. Stream A_i, consisting of requests made of the ith host by remote hosts, competes with local requests for the attention of the request server.

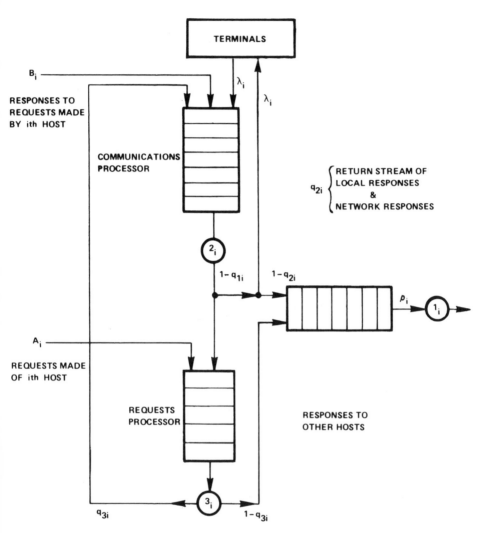

Fig. 20. Detailed model of ith host. (Based on Babic *et al.* [BABI77, p. 74].)

In the event that local requests must be satisfied remotely, they are directed to the loop, which is represented as server 1, operating at rate μ_{1i}. This stream competes for the loop with responses to stream A_i (i.e., responses to requests made of the host). These two streams are merged to form C_i and delivered to the loop. The remote responses to local requests are then returned (after processing) as part of stream B_i to the communications server, where they undergo postprocessing. After postprocessing, the responses are sent to the appropriate terminals.

The model just described fits neatly into the category of a Jacksonian open network of queues. Such a network consists of a collection of (N) nodes with associated queues, each of which behaves as an independent $M/M/1$ queue (i.e., exponential interarrival and service times with one server). The nodes in our case will correspond to host computers. The input rate d_i to node i may be calculated according to the equation

$$d_i = b_i + \sum_{j=1}^{N} r_{ji} d_j \quad \text{for} \quad i = 1, \ldots, N,$$

where b_i is the external arrival rate to node i (in our case, this corresponds to the terminals associated with the host) and r_{ji} the probability of a message being sent from node j to node i. An informative summary of networks of queues is presented by Kleinrock [KLEI75a].

Calculation of Design Parameters

The flow diagram of Fig. 21 indicates the approach described by Babic *et al.* [BABI77] toward the calculation of the design parameters. The following assumptions and definitions will be employed in our calculations:

1. *Assumptions*   The following quantities are assumed known.

(a)   It is assumed that a fractional traffic matrix F, defined as follows, is given:

$$F = \begin{vmatrix} 1 - \sum_{j=2}^{N} f_{ij} & f_{12} & \cdots & f_{1N} \\ f_{21} & 1 - \sum_{j=1}^{N} f_{2j} \ (j \neq 2) & \cdots & f_{2N} \\ \cdot & \cdot & & \cdot \\ \cdot & \cdot & & \cdot \\ \cdot & \cdot & & \cdot \\ f_{N1} & f_{N2} & \cdots & 1 - \sum_{j=1}^{N-1} f_{Nj} \end{vmatrix}$$

The $f_{ij} (i, j = 1, \ldots, N)$ represent the fraction of requests generated at the ith host *which must be satisfied at the jth host.*

(b)   The Poisson arrival rate λ_i, ($i = 1, \ldots, N$) of requests from terminals associated with the ith host , is sometimes known.

(c)   The exponential service rates μ_{2i} and μ_{3i}, ($i = 1, \ldots, N$) for the request and communication server (servers 2 and 3), are assumed known.

(d)   The mean exponential message length $1/\gamma_i$ $(i = 1, \ldots , N)$ is assumed known.

2. *Definitions*   From these quantities, one can calculate the following parameters:

(a) F_i $(i = 1, \ldots , N)$, the fraction of requests generated at the ith host which are to be satisfied remotely.

$$F_i = \sum_{j=1, \, j \neq i}^{N} f_{ij} \, .$$

(b) σ_i $(i = 1, \ldots , N)$, the arrival rate of requests to host i from remote sources:

$$\sigma_i = \sum_{k=1, \, k \neq i}^{N} f_{ki} \lambda_k \, .$$

With these definitions, the authors are able to proceed with the first state—calculation of the transition probabilities.

They first note that the input to server 2 (request server) of the ith host consists of four streams:

1. requests from terminals to be satisfied remotely, arriving at rate $F_i \lambda_i$;
2. requests from terminals to be satisfied locally, arriving at rate $(1 - F_i)\lambda_i$;
3. responses returning to remote requests made by host i at rate $F_i \lambda_i$;
4. responses to local requests returning at rate $(1 - F_i)\lambda_i$.

Referring to the detailed model of the ith host, it is clear that only stream 2 goes to the third server of host i. Hence,

$$q_{1i} = \frac{(1 - F_i)\lambda_i}{F_i \lambda_i + (1 - F_i)\lambda_i + F_i \lambda_i + (1 - F_i)\lambda_i} = \frac{1 - F_i}{2} \, .$$

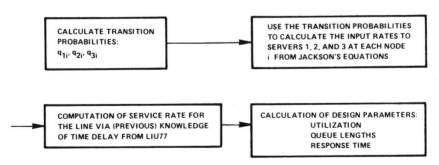

Fig. 21.   Calculation of design parameters.

To calculate q_{2i}, we note that

$$q_{2i} = \frac{\text{stream 2} + \text{stream 3}}{\text{stream 1} + \text{stream 2} + \text{stream 3}}$$

$$= \frac{(1 - F_i)\lambda_i + F_i\lambda_i}{F_i\lambda_i + (1 - F_i)\lambda_i + F_i\lambda_i}$$

$$= \frac{1}{1 + F_i},$$

where we liberally interpret streams 1–3 as the arrival rates of these streams.

To calculate q_{3i}, we note that the input to the request server (server 3) consists of stream 2 and stream A_i.

With the transition probabilities in hand, we now calculate the input rates to the three servers of the ith host, d_{1i}, d_{2i}, and d_{3i}, as

$$\begin{bmatrix} d_{1i} \\ d_{2i} \\ d_{3i} \end{bmatrix} = \begin{bmatrix} 0 \\ (1 + F_i)\lambda_i \\ \sigma_i \end{bmatrix} + \begin{bmatrix} 0 & (1 - q_{1i})(1 - q_{2i}) & 1 - q_{3i} \\ 0 & 0 & q_{3i} \\ 0 & q_{1i} & 0 \end{bmatrix} \begin{bmatrix} d_{1i} \\ d_{2i} \\ d_{3i} \end{bmatrix}.$$

Solving this set of equations, we obtain

$$d_{1i} = \sigma_i + F_i\lambda_i,$$

$$d_{2i} = 2\lambda_i,$$

$$d_{3i} = \sigma_i + (1 - F_i)\lambda_i.$$

This analysis assumes that the structure of each of the hosts is similar—that each host consists of a request server and a communications front end. If the structure of each host differs, the number of equations will increase.

Computation of Loop Service Rate

As already mentioned, an analytical model of the time delay in the loop subnetwork was developed by Liu *et al.* [LIU77]. Using the results of this analysis, an expression for the service rate of the loop as seen by the ith host can be obtained by substituting this expression in the formula for time delay in an $M/M/1$ queue [KLEI75a, p. 98]. This yields the expression for the loop service rate μ_{1i},

$$\mu_{1i} = \frac{1}{T_L} + d_{1i},$$

where T_L is the time delay for the loop as derived by Liu *et al.* [LIU77] and d_{1i} the arrival rate.

Calculation of Design Parameters

The last box in Fig. 21 refers to calculation of the three fundamental design parameters:

1. utilization of request and communication processors as well as the loop communication channel;
2. queue lengths at each of these processors;
3. response time, defined as the time between the arrival of a request from the terminal at the local communication server and the delivery of a response to the terminal.

We approach these calculations as follows:

1. *Utilization*   For processors 2 and 3 (request and communication server), the utilizations are calculated as

$$U_{ij} = \frac{d_{ij}}{\mu_{ij}}$$

for $i = 2, 3$ and $j = 1, \ldots, N$. The utilization of 1 (loop) is calculated as by Liu *et al.* [LIU77]. The calculation will be described later.

2. *Queue lengths*   The average number of messages at the ith server of the jth host, N_{ij}, is given by

$$N_{ij} = \frac{d_{ij}}{\mu_{ij} - d_{ij}}$$

for $i = 1, 2, 3$ and $j = 1, \ldots, N$.

3. *Response time*   The average queueing time T_{ij} at the ith server of the jth host is given by

$$T_{ij} = \frac{1}{\mu_{ij} - d_{ij}}$$

for $i = 2, 3$ and $j = 1, \ldots, N$. The T_{1i} has been calculated before as T_L, for all i.

Using this formula, we can calculate the average response time for requests originating at the ith host to be satisfied at the jth host denoted by T^{ij}.

The components of T^{ij} consist of five time delays:

1. T_{2i}, the mean preprocessing time at the ith host;
2. T_L, the mean loop delay time to deliver the request from host i to host j;
3. T_{3j}, the mean processing time at host j;

4. T_L, for the return trip from j to i;
5. T_{2i}, the mean postprocessing time at host i.

Hence,

$$T_{ij} = 2(T_{2i} + T_L) + T_{3j}.$$

With this expression in mind, T^i, the average response time for remote requests from host i, may be calculated as

$$T^{(i)} = \frac{1}{F_i} \sum_{j=1,\, j \ne 1}^{N} T^{(ij)} f_{ij}.$$

An expression for T, the average response time for any host, may also be obtained:

$$T = \sum_{i=1}^{N} F_i \lambda_i T^{(i)} \Big/ \sum_{i=1}^{N} F_i \lambda_i$$

By similar reasoning, expressions for the average response time to satisfy a local request at the ith host—the system-wide response time for a local request—may be obtained. Denoting these expressions by D^i and D respectively, we obtain

$$D^i = 2T_{2i} + T_{3i}$$

and

$$D = \sum_{i=1}^{N} (1 - F_i) \lambda_i D^{(i)} \Big/ \sum_{i=1}^{N} (1 - F_i) \lambda_i$$

The average response time to satisfy any request from the ith node, $TD^{(i)}$, is given by

$$TD^{(i)} = F_i T^{(i)} + (1 - F_i) D^{(i)}.$$

Hence the average system-wide response time is given by

$$TD = \sum_{i=1}^{N} \lambda_i TD^{(i)} \Big/ \sum_{i=1}^{N} \lambda_i.$$

These analytical results have been compared with results obtained from a GPSS simulation of the network. (Details are given by Reames and Liu [REAM76].)Six hosts were attached to the loop, each host in turn having 50 terminals attached to it. Times and message lengths were assumed to be exponentially distributed. The capacity of the communication channel was varied (10, 20, and 50 kbit/sec).

In order to compare the results of the analysis with the simulation, several parameters pertaining to loop performance were singled out and compared with the simulation results. The parameters were

1. processor and channel utilization and
2. mean system response time and mean response time for a remote request.

Each of these parameters was plotted as a function of the capacity of the loop C and as a function of the fraction of requests to be satisfied remotely (already designated F). The channel and processor utilizations were reported to be in good

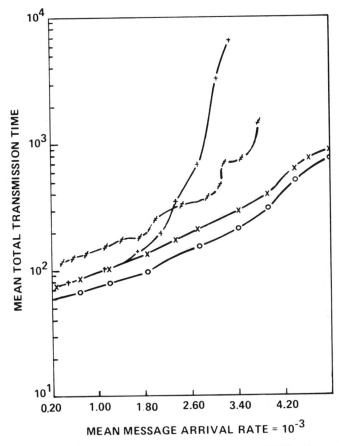

Fig. 22. Mean total transmission time for all three networks. ○, DLCN without acknowledgment time; ×, DLCN with acknowledgment time; ‡, Pierce loop; +, Newhall loop. (Based on Reames and Liu [REAM76, p. 128].)

agreement, as were the response times for high values of C. Some discrepancy did exist for lower values of C probably due to weakness in the formula for T_L.

As part of the same simulation effort, a comparison was also made between the DLCN and the Pierce and Newhall networks [REAM76]. Two of the principal quantities of interest in the comparison were mean total transmission time, defined as time elapsed from message generation to removal of the last character of the message from the loop; and mean queueing time, defined as time elapsed between message generation and its placement on the loop. Figures 22 and 23 portray the dependence of these two quantities on the mean message arrival rate. (A unit of time is equal to the amount of time required to transmit one character.) Both graphs portray the DLCN as superior to either the Pierce or Newhall networks.

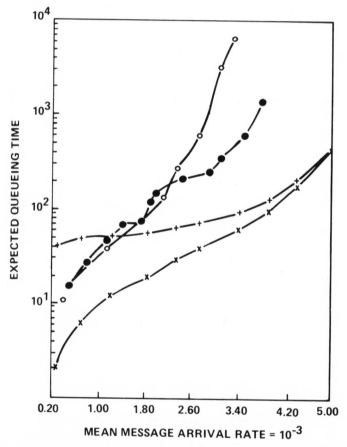

Fig. 23. Expected queueing time for all three networks. ×, DLCN; ●, Pierce loop; ○, Newhall loop; +, DLCN (nonzero). (Based on Reames and Liu [REAM76, p. 128].)

Model for Communications Loop

As mentioned in the preceding section, both analytical and simulation work was done to determine a reasonable expression for the time delay and channel utilization of the DLCN communications loop. The results are reported by Liu *et al.* [LIU77]. With this expression, an expression for the overall time delay for the network was obtained. This overall time delay is defined as the time between the arrival of a request from a terminal at the communications server and the reception of a response at the terminal.

The conceptual model employed for the loop interface is shown in Fig. 24. Three queues are depicted in this model, corresponding to

1. the output buffer of the attached host, which contains messages with a mean arrival rate of λ_i/sec, and a mean message length of $1/\mu_i$;
2. the delay buffer which receives messages arriving at a (mean) rate of γ_i with an average length of $1/\nu_i$;
3. the input buffer which has parameters α_i and $1/\xi_i$.

The fundamental assumptions behind the model of the loop interface are as follows:

1. Both the local data source and the message stream relayed from the input buffer are Poisson processes.
2. The message lengths are taken from a general distribution.
3. The interarrival times of messages and their lengths are independent.

Although neither of the last two assumptions are realistic, they are necessary to make the analysis tractable. The third assumption is the so-called "indepen-

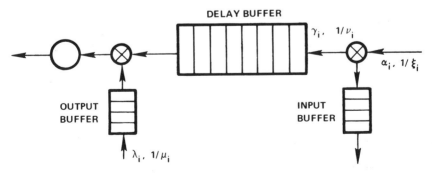

Fig. 24.   Loop interface conceptual model. (Based on Liu [LIU77, p. 73].)

dence'' assumption invoked by Kleinrock in his modeling of the ARPANET [KLEI76, pp. 321–322].

One of the fundamental difficulties in modeling the loop subnetwork is that it is a queueing system in which the priority alternates between two message streams—the local input stream stored in the output buffer (queue 1) and the queue forming in the delay buffer (queue 2). As noted earlier, the incoming messages from the remainder of the loop have priority over locally generated messages until there is sufficient space in the delay buffer to accommodate a local message. At this point, priority switches to the output queue. Mathematically, this condition may be expressed as follows: Messages from queue 1 have priority over messages from queue 2 at time t if and only if the following equation holds at time t:

$$D_i - \sum_{j=1}^{k} m_j \geq S_i,$$

where S_i is the length of the first message in queue 1, k the number of messages in the delay buffer, and m_j the length of the jth message in the delay buffer at time t.

The loop subnetwork itself is modeled as a cyclic network of queues, as illustrated in Fig. 25. The numbers within circles represent channels. Preceding each channel, parameters indicating the average rate(s) of message arrival to and message deletion from the channel are shown. For example, prior to channel i, messages are deleted at a rate of α_i and the arrival at rate λ_i.

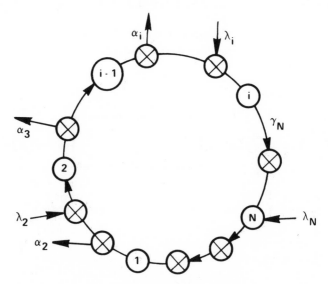

Fig. 25. Loop subnetwork model. (Based on Liu [LIU77, p. 3115].)

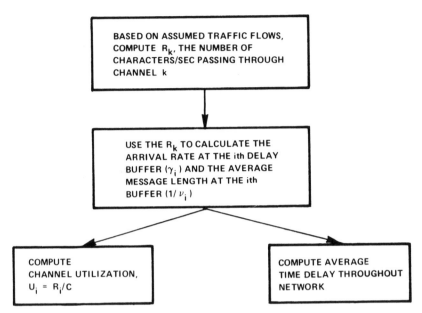

Fig. 26.   Flowchart for parametric calculation.

The second major difficulty in modeling the loop subnetwork is encountered at this point. Because of the almost instantaneous transmission speed of the loop, messages may be *simultaneously* served by two separate channels. For example, the set of characters that comprises a message from node i to node j may be partly in the output buffer preceding node i, and partly in the delay buffer of preceding node $i + 1$. Queueing theory unfortunately assumes that a customer may be served by only one server at a time. Liu *et al.* handle this problem by approximating the alternating nature of the loop service with a nonpreemptive, head-of-the-line priority queueing system. This will be discussed at greater length.

Figure 26 presents an overview of the method set forth by Liu *et al.* [LIU77] for calculating the parameters of network time delay and channel utilization.

Starting at the top of the flow chart, we assume that a traffic matrix P_{ij} is given. Each entry in the matrix represents that traffic originating at node i and destined for node j. The P_{ii} is assumed equal to zero, and $\sum_{j=1}^{N} P_{ij} = 1$.

Our first step is to calculate R_{ik}, the average number of characters per second originating* at node i and passing through node k (on the way to node j, for example).

The average number of characters per second going from node i to node j is given by $P_{ij}\lambda_i/\mu_i$, as the input process is Poisson by assumption.

---

*Node i represents *channel i*, and traffic originates at a computer system attached to channel i.

Hence an expression for R_{ik} can be obtained as

$$R_{ik} = \frac{\lambda_i}{\mu_i} \sum_{j=1}^{i-1} P_{ij} + \frac{\lambda_i}{\mu_i} , \qquad 1 < i < k \neq N,$$

$$= \frac{\lambda_i}{\mu_i} \sum_{j=1}^{i-1} P_{ij} , \qquad 1 < i < k = N,$$

$$= \frac{\lambda_i}{\mu_i} \sum_{j=k+1}^{N} P_{ij} , \qquad 1 = i < k \neq N,$$

$$= \frac{\lambda_i}{\mu_i} \sum_{j=k+1}^{i-1} P_{ij} , \qquad k + 1 < i \leq N,$$

$$= \frac{\lambda_i}{\mu_i} , \qquad i = k,$$

$$= 0, \qquad \text{otherwise.}$$

These expressions are best understood by referring to Fig. 25. R_k can now be calculated as

$$R_k = \sum_{i=1}^{N} R_{ik} , \qquad k = 1, \ldots , N.$$

Proceeding to the second box in Fig. 26, we calculate the parameters for the delay buffer, γ_i and $1/\nu_i$, as (these expressions are useful in calculating the average message delay)

$$\gamma_i = \sum_{j=1, \, j=i}^{N} \mu_j R_{ji} .$$

This expression follows from the definion of R_{ji} as equal to $P_{ji} \lambda_j / \mu_j$, since $\mu_j P_{ji} \lambda_j / \mu_j = \lambda_j P_{ji}$, and the summation of these expressions must total γ_i, the input to the delay buffer.

To calculate $1/\gamma_i$, the authors first define 1_{ji} to be that part of the traffic passing through delay buffer i which originates at node j. Then,

$$l_{ji} = \begin{cases} R_{ji} \, / \, (R_i - \lambda_i / \mu_i) & \text{for } j \neq i, \\ 0 & \text{for } j = i \end{cases}$$

and

$$\frac{1}{\gamma_i} = \sum_{j=1}^{N} \frac{l_{ji}}{\mu_j}$$

Thus, $1/\gamma_i$ is a traffic-weighted average message length.

Proceeding to the third level of Fig. 26, we calculate the channel utilization at node i via the equation

$$U_i = \frac{R_i}{C}.$$

Finally, we calculate the average message delay in the network. Our first step is to obtain an expression for the average time delay T_{ij} between nodes i and j. There are five components in such a delay:

1. waiting time in the delay buffer (queue 1), denoted by $T_1^{(i)}$;
2. time required to multiplex the message onto the loop, denoted by $T_2^{(i)}$; this time is simply equal to M/C, where M is the length of the message and C the capacity of the line;
3. time T_3 required to check the address field of the message header at each intermediate node, equal to B/C sec, where B is the number of characters in the address field;
4. waiting time in each of the k intermediate delay buffers, denoted by $T_4^{(i+k)}$;
5. propagation delay for the network, T_5; T_5 is negligible for a local network, and is therefore set equal to zero ($T_5 = 0$).

If we assume that there are $r - 1$ intermediate nodes between node i, the source, and node j, the sink, then we have the following expression for T_{ij}:

$$T_{ij} = T_1^{(i)} + T_2^{(i)} + rT_3 + \sum_{k=1}^{r-1} T_4^{(i+k)}$$

$$= T_1^{(i)} + \frac{M}{C} + \frac{rB}{C} + \sum_{k=1}^{r-1} T_4^{(i+k)}.$$

To find the average message delay, we take expectations of both sides of this equation, and arrive at the formula

$$E(T_{ij}) = E(T_1^{(i)}) + \frac{1}{\mu_i C} + \frac{rB}{C} + \sum_{k=1}^{r-1} E(T_4^{(i+k)}).$$

Note that $E(m) = 1/\mu_i$ in this formula.

To get an expression for the average time delay T in the network, we apply Little's law [KLEI75c, p. 17] to the preceding formula, and obtain

$$T = \sum_{i=1}^{N} \left[E(T_1^{(i)})\lambda_i + \frac{\lambda_i}{\mu_i C} + E(T_4^{(i)})\gamma_i \right] \bigg/ \sum_{i=1}^{N} \lambda_i + E(r)\frac{B}{C},$$

where $E(r)$ is the average path length, an expression for which is

$$E(r) = \sum_{i=1}^{N}(\gamma_i + \lambda_i) \Big/ \sum_{i=1}^{N}\lambda_i$$

(see [KLEI76, pp. 119–28]). Expressions for $E(T_1^{(i)})$ and $E(T_4^{(i)})$ are missing from this discussion. It is in providing expressions for these two expectations that Liu and his associates provide their approach to the problem of the alternating priorities of their queueing structure. They assume that locally generated messages always have priority over incoming messages, *a situation that exists in low-traffic conditions* (if the size of the delay buffer is large enough to hold any message generated at the node).

Given the assumption, the queueing structure fits neatly into the category of a nonpreemptive, head-of-the-line priority queueing system [KLEI76, pp. 119–123], for which the following formulas apply:

$$E(T_1^{(i)}) = \frac{W_0}{(1 - \rho_2)} \, ,$$

$$E(T_4^{(i)}) = \frac{W_0}{(1 - \rho_1)(1 - \rho_2)},$$

where

$$W_0 = \frac{\lambda_i \overline{a_i^2}}{2C^2} + \gamma_i \frac{\overline{d_i^2}}{2C^2} \, ,$$

$$\rho_1 = \frac{\lambda_i}{\mu_i C} + \frac{\gamma_i}{\nu_i C} \, ,$$

and

$$\rho_2 = \frac{\lambda_i}{\mu_i C} \, .$$

A GPSS simulation was written to verify the assumptions of the analytical model, the primary quantities of interest being channel utilization and average message delay. The results of the simulation indicated good agreement on channel utilization. Under low traffic conditions (corresponding to a utilization of at most 0.3 or 0.4), the average message delays are also close. As traffic increases, agreement decreases as the analytic models provide more conservative results for the time delays. As mentioned by Liu *et al.* this discrepancy is no doubt due to

1. assumption of Poisson arrival rate at the delay buffer,
2. assumption of independence of message lengths and message interarrival times,
3. approximation of the alternating priority structure by a fixed structure.

DDLCN

The DDLCN (double distributed loop computer network) is a continuation of the DLCN program at Ohio State University, under Professor M. Liu.

Wolf [WOLF79a] describes the fundamental aim of the design as being the creation of a fault-tolerant ring network with good performance characteristics.

Wolf's motivation was simple—the loss of a single link on a ring network results in a complete loss of service as either a message or its acknowledgment must be blocked.

As a solution to this problem, Wolf employs two separate loops transmitting in opposite directions in his design. Each of the loops employs the DLCN shift register insertion technique. In the event of a failure, the loop interface unit (LIU) reroutes messages from one loop to another, by-passing the failed portion of the loop.

In addition to achieving a degree of fault tolerance, the design for the DDLCN exhibits a performance superior to that of the DLCN.

Our discussion of the DDLCN is divided into three sections. The first section describes the operation of the network, including a description of how the network reconfigures itself in the event of link failures. The following two sections are devoted to the description of an analytical and a simulation model of the performance of the networks in the presence and absence of faults.

Description

Work done by Zafiropulo on the reliability of loop networks [ZAFI74] had, according to Wolf, an important influence on the design of the DDLCN.

Zafiropulo's ideas on the reliability of ring networks have also influenced other designs, for example, the Litton-built distributed processing system [MAUR79]. Consequently, we feel that it is worthwhile to briefly summarize some of his major ideas.

Zafiropulo [ZAFI74] suggests three techniques for improving the reliability of a ring network. All three involve the addition of a second loop to the original loop network. In the event of a link failure, messages are routed onto the second loop around the failed portions of the original loop.

In the failure by-pass method, pictured in Fig. 27, both loops transmit messages in the same direction. In the event of a failure, a portion of the stand-by loop is employed to conduct messages around the failed portion and back onto the main loop.

In the self-heal technique, the stand-by loop transmits messages in the opposite sense from the original loop. As pictured in Fig. 27, the main and stand-by

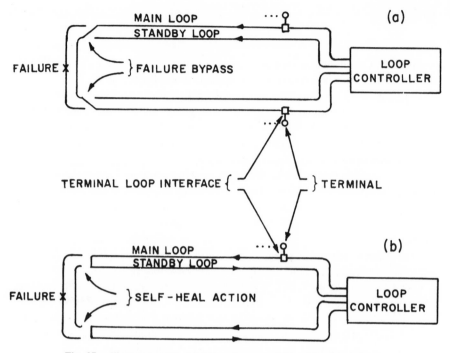

Fig. 27. Illustration of the (a) failure bypass and (b) self-heal techniques.

loops join one another to form a new loop, thereby excising the failed portion of the loop. This technique is employed in the Litton-built distributed processing system (DPS) which is described by Mauriello *et al.* [MAUR79]. Reliable operation is extremely important to the DPS, as it is intended as a command and control network.

Wolf adopted from Zafiropulo the idea of employing two loops in the network, which transmit in opposite directions. However, instead of using one of the loops as a stand-by (as in the DPS), he uses both loops to transmit data *simultaneously.* Figure 28 depicts this notion.

Each loop makes use of the shift-register insertion technique employed by the DLCN to transmit its messages on the loop(s). The link interface unit (LIU) at each node is also responsible for choosing the loop on which to route a message, making its decision on the basis of a shortest distance algorithm. A master link map (MLM) and a master host map (MHM) are kept in each node, indicating the status of all of the links and nodes in the network. If a node wishes to send a message, it first checks the status of the destination host to see if it is capable of accepting messages. If it is, it then checks the status of the links in the network, and determines the shortest path over which to route the message to its destination.

In order to implement this mode of operation, each of the LIUs is provided

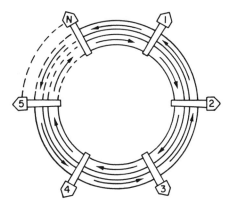

Fig. 28. The two loops of the DDLCN.

with tristate logic, driven by a microprocessor. A functional design for the LIU is pictured in Fig. 29.

Fault-Present Operation

As already indicated, the basic object of the design was to develop as high as possible a level of fault tolerance in the system in the presence of link outages.

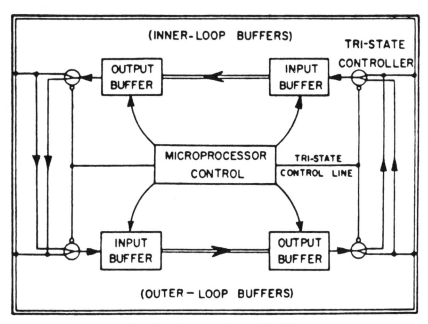

Fig. 29. Functional design of the LIU.

The strategy for handling either LIU or host failures is to regard the links emanating from the failed node as having failed.

In the event of a failure on a single link, the basic objective of the DDLCN *is to keep one of the two loops logically intact*. The particular loop to be kept alive is picked when the network is first brought up.

Each LIU on the loop continually monitors its upstream lines for the presence of either messages or timing signals. Timing signals are broadcast by an LIU where there is no message to send. The absence of any signals on an upstream line indicates that the line is down.

In the event that a link fails on the expendible loop, the network merely broadcasts a message indicating that the MLM must be updated. If a failure occurs on the loop which is to be kept alive, the LIU first reverses the direction of transmission on the partner of the failed link and then routes messages over that link by connecting the input buffer leading to the failed link with the output buffer of its replacement.

Both of these schemes are depicted in Fig. 30. In this figure, the inner loop is

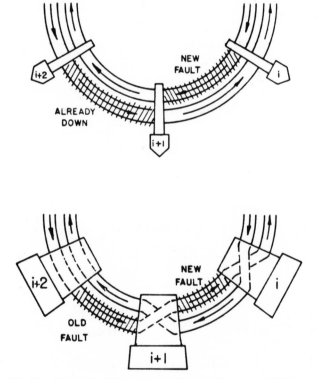

Fig. 30. Portrayal of link exchange. (From Wolf *et al.* [WOLF79b].)

picked as being the loop to be maintained. A fault occurs on the outer loop, resulting in a message updating the MLM being broadcast to the remaining nodes. This is followed by a failure occurring on the inner loop, in which case messages are rerouted onto the partner of the failed loops. The inner loop is thus kept logically, albeit not physically, alive.

A double-fault condition occurs when both partners of a link pair are broken. If there are no single faults present on the remaining links, communications among all nodes still exist. If, however, there are single faults on the remaining links, the network uses two transmission mechanisms in order to be able to transmit messages in *both directions*. In Fig. 31, if node 4 wishes to send a message to node 5, the direction of message flow between nodes 4 and 5 must be reversed. In order to accomplish this, Wolf employs the Newhall protocol in addition to the DDLCN protocol in the following fashion.

The LIU which detects a double fault first broadcasts a message to the remaining nodes on the network informing them of the double failure, and then generates a Newhall control token. The token travels around the loop until it reaches the other side of the double fault. It then reverses its direction and proceeds to bounce back and forth from one side of the double fault to the other. After passing over a link, it reverses the direction of transmission over the link, thus allowing transmission in the opposite direction to that of its travel.

If a node wishes to transmit a message (e.g., node 4 wishes to send a message to node 5), it first examines the MLM to see if a path is available to the destination node. If a path is available, the DDLCN protocol is employed to send the message. If no path is available, the LIU waits for the control token to arrive and

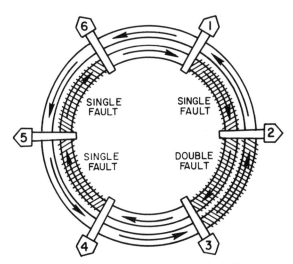

Fig. 31. One double fault and three single faults.

then sends its messages via the Newhall protocol. In order to send a message via the Newhall protocol, the direction of message flow must be reversed during the preceding trip of the token. In our example, if the token were traveling from node 3 to node 4, the message flow along the link between nodes 4 and 5 had to be reversed before the token reached node 4 in order to enable node 4 to send its message to node 5.

Reversing the transmission direction on a link once the token has passed over the link also has the benefit of allowing a node to send all of its messages once it possesses the control token, regardless of their destination. In our example, node 4 can send a message to node 3 as well as to node 5 once it has the token, since the direction of message flow between nodes 3 and 4 is reversed once the token traverses it on the way to node 4.

Further double faults break the loop into disjoint, noncommunicating sections. However, communication among nodes within those sections can be maintained by the foregoing methods. For more details on the operation of the loops, the reader is advised to consult Wolf [WOLF79a or WOLF79b].

We now continue with a discussion of the analytical and simulation models of the DDLCN.

Queueing Model

The queueing model developed by Wolf* avoids the difficulty encountered by Reames in portraying the DLCN transmission mechanism (i.e., the alternating priorities of the queues for access rights to the line) by not attempting to solve the problem. Instead of developing an expression for the queueing delay at the LIU, Wolf represents the DDLCN as a closed queueing system, modeling the traffic flow at each LIU as an $M/M/1$ queue, and invoking Jackson's theorem [KLEI75a] to calculate message arrival rates at the LIUs.

In doing so, an accurate model of the time-delay performance of the network is no longer possible. It is, however, possible to calculate link utilizations as a function of message arrival rates and routing probabilities.

Accordingly, *the major goal of the queueing model was to determine the traffic levels which correspond to the network saturating* in both the presence as well as the absence of link failures. The results were compared to the performance of the DLCN.

Figure 32 presents a model of the message flow across the LIU. Definitions of the symbols used in the figure are as follows:

---

*The reader is assumed to have read the section on the modeling of the DLCN before proceeding with this section.

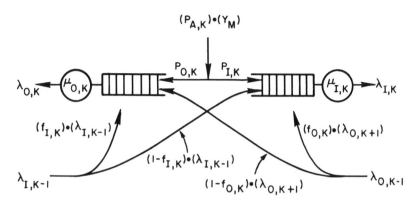

Fig. 32. Queueing model for the LIU. (From Wolf [WOLF79a].)

γ_M total rate at which messages are generated by network nodes

$P_{A,k}$ proportion of all messages originating at node k; $\sum_{k=1}^{n} P_{A,k} = 1$

$\mu_{I,k}$, $\mu_{O,k}$ service rate of transmitter for (inner, outer) loop for node k

$P_{I,k}$, $P_{O,k}$ probability that message is sent out over the (inner, outer) loop from node k; ($P_{I,k} + P_{O,k} = 1$)

$\lambda_{I,k}$, $\lambda_{O,k}$ departure rate of messages over (inner, outer) loop from node k

$f_{I,k}$, $f_{O,k}$ fraction of messages on (inner, outer) loop reaching node k that get off their respective loop at node k

Computation of the message arrival rate which saturates a link is a straightforward matter. First, it is necessary to calculate the arrival rates at the LIUs, given the (outside) message arrival rates as well as the routing probabilities (assuming shortest path routing). This expression will depend on the total message arrival rate to the system, once the other parameters are known. This expression is then equal to the transmission rate of the LIU and the resulting equation is solved for the total message arrival rate as a function of the message transmission rate.

An example of this computation for a six-node homogeneous network is portrayed in Fig. 33.

A comparison of the message arrival rates which produce saturation was made between the DDLCN and DLCN networks. Completely homogeneous traffic statistics were assumed in both cases (i.e., all hosts generate messages at the same rates, with the same length distributions, and send them to other hosts with equal probabilities). By employing the approach just outlined, it was discovered that

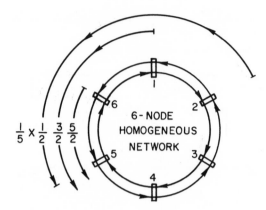

Fig. 33. Link saturation in the DDLCN. $\lambda = (4\frac{1}{2}/5)(\frac{1}{6}\gamma_M)$; for saturation: $\lambda = \mu = \frac{9}{10}(\frac{1}{6}\gamma_{M\ SAT})$; thus, $\gamma_{M\ SAT} = (\frac{20}{3})\mu$. (Based on Wolf [WOLF79b].)

1. the DLCN saturates at twice the nodal transmission rate, independent of the number of nodes n in the network; and

2. the DDLCN saturates at $8(n - 1)/n$ times the transmission rate if n is even and at $8n/(n + 1)$ times the transmission rate if n is odd.

Hence, the (homogeneous) DDLCN saturates at a traffic level which is $4n/(n + 1)$ times that which would saturate the (homogeneous) DLCN, if n is odd, and $4(n - 1)/n$ times the DLCN saturation rate for n even.

Clearly, as n increases the foregoing factor approaches 4. This factor of 4 is basically due to the doubling of the bandwidth in the DDLCN as well as the fact that a message travels, on the average, only one-fourth of the way around the loop in the DDLCN versus one-half of the way around in the DLCN.

Graphs depicting the link utilization under fault-present conditions were developed by Wolf for a number of homogeneous as well as nonhomogeneous networks. For a discussion of some of the phenomena which were observed, the reader is referred to Wolf [WOLF79a,b].

Simulation Model

The purpose of the simulation model was to fill in the gaps left by the analytical model as well as to confirm the results obtained by the analytical model.
The simulation was used to investigate

1. message queueing, propagation, and acknowledgment times;
2. total transmission time, equal to the sum of the above three times;

3. delay–buffer queue lengths;
4. link utilizations; and
5. message routing strategies other than shortest distance.

Values for these quantities were obtained under various conditions of link fault conditions and message input rates.

As the model was quite complicated, a partial factorial design* was necessary to determine gross network performance characteristics before preceding with a full-factorial design. The following parameters were then assumed to be fixed:

1. number of nodes at 6;
2. data messages exponentially distributed with a mean length of 50 characters, and acknowledgments at 6 characters;
3. transmitter service rate at 100,000 characters/sec;
4. 512-character delay buffer;
5. transmission error rate of 1 in 10,000;
6. shortest distance routing scheme; and
7. uniform message addressing probabilities.

The simulation program was written in GPSS and was implemented on an Amdahl 470. The results indicate a close agreement with the link utilizations predicted by the analytical model.

The message-delay performance is depicted in Fig. 34, which portrays "relative transmission time" versus "no-fault utilization."

The no-fault utilization is the utilization in a network with no faults. The relative transmission time is the ratio of the average message transmission time to the best possible transmission time, that is, the transmission time of a message in the absence of other message traffic.

Two interesting facts pointed out by the figure are:

1. There is not much of a difference between the performance of the network with one single or one double fault. They both overload at about one-half the arrival rate for a fault-free network.

2. As predicted by the analytical model, the DDLCN performance is significantly better than the DLCN performance. The *DLCN saturates at approximately 30% of the traffic level which the DDLCN supports.*

---

*A factorial design of a (simulation) experiment is a method of varying several parameters simultaneously to avoid excessive runs while, at the same time, maintaining statistical validity of the results. See Fishman [FISH73].

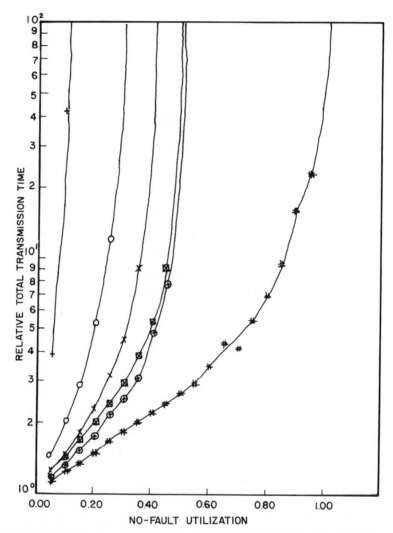

Fig. 34. DDLCN response time curves: +, 1 double and 5 single loops; ○, single loop only; ×, 2 single faults; ⊗, 1 double fault; ⊕, 1 single fault; #, no faults. (Based on Wolf [WOLF79b].)

Another major area of interest in the simulation studies was an investigation of the effects of routing policies other than shortest-distance routing on message transmission times.

Assuming a "homogeneous" network, two basic policies were investigated. Both of the policies involve maintaining statistics on message traffic at each node in the network in an attempt to balance the traffic on both loops. The two policies were:

1. Keep track of the elapsed time for the last k messages at each node, and select the direction to route a message by minimizing the expression

$$\frac{\text{distance that a message travels on the loop}}{\text{total elapsed time for the last } k \text{ messages on that loop}}.$$

This policy penalizes the loop with more traffic.

2. Keep track of the elapsed time for the last k messages as well as the number of characters which pass by, and choose the direction to route a message by minimizing the foregoing expression *multiplied by the number of characters in the last k messages.*

These algorithms attempt to distinguish between a long elapsed time produced by long messages and a long elapsed time produced by a collection of shorter messages.

The simulation results indicate that these two algorithms do not improve on the performance of the shortest-distance routing algorithm. For further details on the investigation of routing algorithms, the reader is referred to the work of Wolf [WOLF79a,b].

OREGON STATE LOOP

Jafari *et al.* [JAFA78] describe a design for a double-loop communications network. One of the loops is used for control packets, while the other is employed for transmission of data packets. As pictured in Fig. 35, the loop makes use of a central controller to assign communication paths on the data loop to requesting nodes.

Thus, for example, if node 5 wishes to speak to node 2, it first sends a control packet via the control loop to the controller requesting a connection to node 2. If a connection is available, the controller then sends messages to each of the nodes along the path between nodes 2 and 5 to close the appropriate switches, thereby forming a path for data flow. The controller also informs node 2 of an impending message.

The role which the loop supervisor plays in the operation of the loop clearly *prevents control from being completely distributed.* Nevertheless, simultaneous communication between nodes is possible, and the loop does have impressive performance characteristics, as will be discussed.

Both simulation and analytical studies on the proposed loop were conducted. The simulation studies compare the performance of the loop to the performance of the Pierce, Newhall, and DLCN loops, and indicate that the proposed Oregon State loop has a superior delay versus throughput profile.

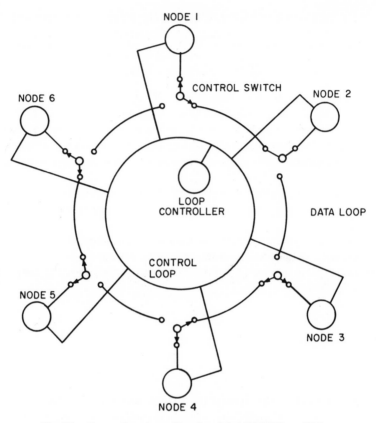

Fig. 35. Oregon State loop. (Based on Jafari [JAFA78, p. 72].)

The analytical studies done on the loop are interesting, as the authors propose an experimental approach toward the development of analytical performance models. They first simulate the loop, thereby developing performance data, and then curve-fit the data *with queueing formulas* instead of the usual linear or polynomial forms dictated by numerical analysis.

The following two sections discuss both modeling approaches, as well as the conclusions derived by them.

Comparison with the Pierce, Newhall, and DLCN Loops

Jafari *et al.* based their comparisons of these four architectures on results obtained by Reames and Liu [REAM76] from simulation studies comparing the Pierce, Newhall, and DLCN loops. After validation of Reames' results, Jafari used Reames' figures comparing the three loops.

A brief list of the assumptions of the model, including those imposed by Reames, are the following:

1. Message lengths are exponentially distributed, with a mean length of 50 characters, and can vary between 10 and 512 characters.

2. Message arrivals are governed by a Poisson distribution.

3. An optimal packet size of 36 characters for the Pierce loop is selected and obtained by minimizing the product of the average number of packets per message by the packet size.

4. The Newhall network was simulated so that the control token was passed by a node only when the message queue at the node was empty.

5. Performance results for the DLCN, Newhall, and Pierce loops are confined to a six-node network, while the new loop was studied for a number of nodes ranging from 2 to 15.

Figure 36 portrays the mean total transmission time versus throughput results obtained by the simulation for a six-node network. The mean total transmission time for all of the loops *except the DLCN* is defined as the sum of the queueing time and the transmission time. The transmission time for the DLCN includes the message acknowledgment times. As indicated in the figure, the performance of four separate configurations of the Oregon State loop are compared to the performance of the Newhall, Pierce, and DLCN loops. The simplest of the four, a single half-duplex loop operating under a first-in first-out scheduling discipline, is depicted as having superior performance characteristics to those of its three competitors. The figure also indicates improvements in the performance of the loop obtained by other configurations, such as the addition of a second data loop.

Details of the simulation studies are presented by Jafari [JAFA77].

The *M/M/Z* Approximation

As already indicated, Jafari's approach toward developing an analytical model for the loop was essentially experimental. He first simulated the response time versus throughput behavior of the loop, and then employed queueing formulas to fit the data.

The first formula which he attempted to use had the form

$$T = K_1 + \frac{K_2 t_s}{1 - \lambda t_s / K_3} ,$$

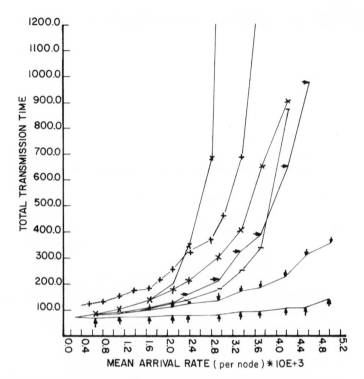

Fig. 36. Comparison of response times for four variations of the Oregon State loop. *, Newhall loop; +, Pierce loop; ×, DLCN; new loops: →, one way, one data loop, first-in, first-out; -, one way, one data loop, shortest message first; ↓, two way, one data loop, FIFO; ↑, one way, two data loop, FIFO, per loop. (Based on Jafari [JAFA78, p. 75].)

where K_1 represents all "reasonably" constant system delays (i.e., low-load delays) and the second term represents the saturation effects in the system, λ is the arrival rate and t_s is the service time.

One of the motivations for using this formula was the fact that the Pollaczek–Khinchin formula [KLEI75a] for the time delay in the $M/G/1$ system has (approximately) this form.

The constant K_3 can be interpreted as the number of servers in the system. Attempting to fit the preceding formula to the simulation data, Jafari noted nonintegral values for K_3, with $0 \leqq K_3 \leqq 1$. Jafari noted that interpreting K_3 as the number of servers was still reasonable, in light of the fact that *the number of connections in the loop, that is, the number of servers, varies over a period of time.*

As a result of this heuristic chain of reasoning, Jafari was led to seek a gen-

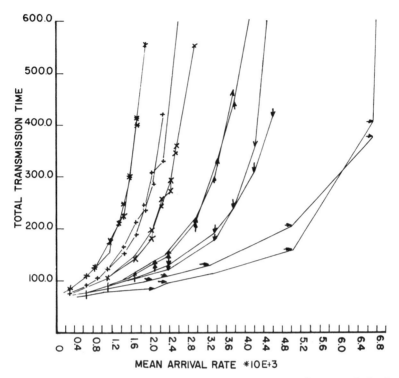

Fig. 37. Comparison of queueing and simulation results: →, 2 nodes; ↓, 4 nodes; ↑, 6 nodes; ×, 9 nodes; +, 11 nodes; *, 15 nodes. (Based on Jafari [JAFA78, p. 76].)

eralization of the time delay formula for an $M/M/m$ system in which the number of "servers" could be nonintegral, lying in between 0 and 1.

The formula for time delay in an $M/M/m$ system is given by

$$T = t_s + \frac{P_m t_s}{m(1 - \rho)} ,$$

where P_m is the probability that an arriving customer has to queue, t_s the mean service time, m the number of servers, and ρ the utilization $\lambda t_s/m$.

Noting that it was only necessary to generalize P_m, Jafari did so, and developed the expression

$$P_z = \frac{z\rho^z}{1 + (z - 1)\rho}$$

Since this expression fit the simulated data well, Jafari adopted it in the place of P_m, and christened the result the $M/M/Z$ approximation.

Adding the formula for the time delay on the control loop (modeled as an $M/M/1$ queue) to the $M/M/Z$ expression, he obtained the expression for the overall time delay on the loop;

$$T = t_c + \frac{\rho_c t_c}{1 - \rho_c} + t_d + \frac{\rho_d^z t_d}{(1 - \rho_d)(1 + (z - 1)\rho_d)} .$$

The first two terms represent the control-loop time delay, while the last two represent the delay on the data loop.

As Fig. 37 illustrates, the expression is a good fit for the simulation results. The smoother curves in the figure represent the queueing results.

MODELING OF RING NETWORKS—CONCLUDING THOUGHTS

Our first topic in this final (ring) section will be a comparison of the performance characteristics of loop control architectures, employing our familiar time delay versus throughput curves. The comparison is necessarily limited, however, by the lack of modeling which compares all of these architectures in a uniform fashion, that is, employing the same parameters and assumptions.

The following section contains a discussion of the models themselves, focusing on the difficulties in developing these models, and presenting several suggestions for the modeling of ring networks.

Performance Comparison of Ring Networks

As already mentioned, no comprehensive model has been developed which compares the performance of all of the ring networks which we discussed with respect to a uniform set of parameters and assumptions. Baghdadi, in England, is at work on such a comparative evaluation, but his results are not as yet available.

A number of modeling efforts were conducted by proponents of the various loop architectures in an attempt to compare the performance of their own proposed architecture with that of other rings.

1. The performance of the Oregon State loop was compared to that of the Pierce, Newhall, and DLCN loops. It should be noted that the DLCN results were used as a baseline in performing the comparisons of the networks.

2. The performance of the Waterloo loop was compared to that of the Pierce and Newhall loops.

3. The performance of the DLCN loop was compared with that of the Pierce and Newhall loops. Making use of the same assumptions, the DDLCN was compared to the DLCN.

All of these studies have been described in the preceding pages. Now we attempt to draw whatever conclusions possible in spite of the lack of uniform conditions for making these comparisons.

All of the studies, with the exception of the one done at the University of Waterloo, agree on the comparison of the Pierce and Newhall loops. The Newhall loop shows a better performance at low loads. In the middle throughput range, the performance of the two loops is comparable, while in the higher throughput regions, the Pierce loop exhibits superior performance. Figure 38,

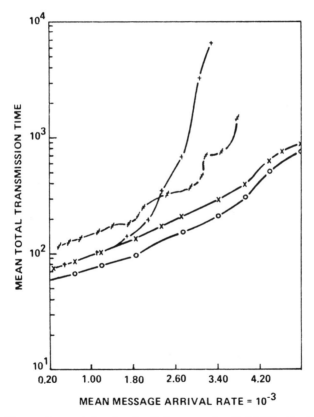

Fig. 38. Mean total transmission time for all three networks. O, DLCN without acknowledgment time; ×, DLCN with acknowledgment time; ‡, Pierce loop; +, Newhall loop. (Based on Reames and Liu [REAM76, p. 128].)

taken from the section on the DLCN, compares the performances of the Pierce, Newhall, and DLCN loops.

An explanation for this behavior is that at the lower loads, a packet will have almost immediate access to the line in the Newhall loop because the control token circulates very quickly. In the Pierce network, a packet must wait an average of half a slot before obtaining access to the line at these same low loads. Once traffic builds up, the ability of the Pierce loop to handle multiple "conversations" asserts itself, and its message delay decreases below that of the Newhall loop.

The Waterloo study portrays the Newhall loop as having superior performance to that of the Pierce loop for all throughput levels. In comparing the various studies, it is not absolutely clear what the reasons are for this difference. One possibility is that the Waterloo study does not appear to have optimized the packet size in an effort to minimize transmission time, whereas other studies have done this.

A point made by the Waterloo study is that if one is willing to pay the price of having control of access to the loop completely decentralized, a significant improvement can be made in the performance of the Pierce loop.

Those users having the longer message queues are granted access to the line. This eliminates the "loop-hogging" behavior characteristic of the Pierce loop, and provides superior performance for nonhomogeneous loads.

Richardson and Yu [RICH79] compare the Pierce and Newhall architectures in the context of a centralized inquiry–response system. They assume a collection of terminals making inquiries of and receiving responses from a central computer. Also included in the study is a variant of the Pierce access mechanism in which a user is not allowed to write into a slot from which he is receiving. The authors refer to the three protocols as

1. pass control (PC) for the Newhall loop,
2. slot deletion (SD) for the Pierce loop, and
3. slot no deletion (SND) for the Pierce loop in which a user cannot write into a slot from which he receives.

Both queueing and simulation models were developed in order to compare the response time of the three systems with one another. The response time is defined as the "elapsed time from inquiry generation at the source terminal until reception of the response, but excluding computer processing time."

Figure 39 contains a comparison of the three disciplines for the following typical operating parameters: $n = 100$ users, ρ (utilization) $= 0.682$, and m_f (inquiry length) $= m_r$ (response length). This figure portrays the effect of message length on time delay in the three networks. For short messages (100–600 bits), the SD (Pierce) loop has a higher delay than the PC (Newhall) loop.

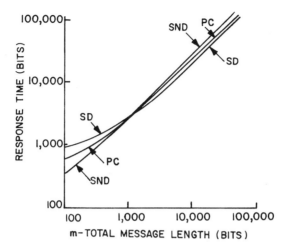

Fig. 39. Comparison of SD, PC, and SND loops ($m_t = m_r$). (Based on Richardson and Yu [RICH79, p. 63].)

For medium-length messages (600–1000 bits), the performance of the two systems is comparable, while for the longer messages, the SD loop is superior for nonhomogeneous loads.

The results of the DLCN modeling as portrayed in Figs. 22 and 23 clearly point to the fact that the DLCN is superior to either the Pierce or the Newhall networks. Both analytical as well as simulation models confirm this. The simulation study described by Reames and Liu [REAM76] points out the superior performance of the DLCN even when message acknowledgment times are included, and when the packet size is optimized in the Pierce network.

The DDLCN, as would be expected, provides superior performance to the DLCN. For example, the DLCN saturates at approximately 30% of the load that the DDLCN can tolerate. More significant than this, however, is the fact that the DDLCN provides a fault-tolerant communications capability, something which previously has been totally lacking in either ring or bus networks.

The Oregon State loop, by relegating control traffic to a separate loop and employing a central controller to close switches between senders and receivers, provides a performance roughly comparable to that of the DLCN. As with the Waterloo loop, one must be willing to pay the price of centralized control for access to the loop.

Performance Models of Loop Access Protocols

This (ring) section has emphasized discussion of queueing models for the various major network architectures. With but one exception [KAYE72], the queueing models were compared to simulations of the system under consideration. Agreement between the queueing model and the simulation was generally good, especially under conditions of moderate line loading. (The general rule-of-thumb definition for "moderate" is 60% utilization.)

Unfortunately, there appears to be no literature comparing these models (simulation or analytic) with actual systems. However, people associated with the DLCN project intend to make such comparisons in the near future, once their experimental DDLCN network is constructed. It is certainly unfortunate that so little work has been done in comparing models and actual systems, as much could be gained from such work.

It appears that the major difficulty in constructing queueing models of ring networks is the modeling of the ring subnetwork. The ring subnetwork must be distinguished from the system, which includes both the ring and the host processors attached to it. This has been a difficulty with all the models considered.

There is no comprehensive time-delay model for the Newhall network. A distribution for time delay has been derived under light traffic conditions [KAYE72], and a time-delay formula developed for the two-host case [LABE77]. Other results on the network are devoted to scan time.

In modeling the Pierce network, Hayes was forced to construct two separate models for different traffic patterns [HAYE71]. The modeling of the DLCN and Waterloo loops encountered difficulties because of the alternating priorities of the queueing network. Jafari, in modeling the Oregon State loop, completely avoided the issue by fitting queueing formulas to his simulation results.

In light of all these difficulties, it appears that a reasonable approach to the modeling of ring networks would be to simulate the ring subnetwork, and to employ a (Jacksonian) open network of queues to represent the nodes. If heavy traffic conditions are not deemed vital to investigate, then the queueing models already discussed should prove to be adequate. As they provide conservative estimates for the time delay under heavy traffic conditions, little will be lost by employing them. In any event, further research in the development of queueing models for ring networks would certainly be of value.

Chapter 3

Bus Networks

INTRODUCTION

The use of a bus as the communications subnetwork is a very popular approach to construction of local networks. Anderson and Jensen [ANDE75] have created a taxonomy of the possible communications systems for a network of computers. They characterize a bus network as having its processing elements (computers, peripherals) attached to a common channel (the bus). Their conception of a bus network is shown in Fig. 40.

The channel itself is employed in a broadcast mode—all processing elements "hear" a message. The transmission medium for the channel can be coaxial cable, optical fibers, twisted pairs, etc. Access to the network is controlled via some time-multiplexing technique.

As with ring networks, the major advantage of a bus network is its simplicity. It is easy to add or delete processing elements since numerous connections do not have to be made with each new addition (or deletion) of an element. In addition, start-up and modification costs are low compared to other types of networks.

The principal disadvantage of a bus network is the vulnerability of the network to a failure of the bus itself. However, processing element failures are not catastrophic. A ring network must overcome this problem by using bypass units at the nodes.

In the case of either a ring or a bus network, some form of redundancy in

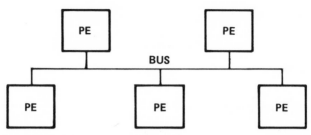

Fig. 40. Bus network; PE, processing element. (Based on Anderson and Jensen [ANDE75].)

the communication channel is necessary to eliminate the vulnerability of the system to a channel outage.

Performance evaluations of bus networks have centered on evaluating the access protocols by which the nodes gain access to the bus. Luczak [LUCZ78] develops an exhaustive classification of techniques employed in constructing bus networks. The bulk of his paper is devoted, in fact, to a description of channel access techniques.

Luczak points out that there are three major categories of access protocols: selection, random access (or contention), and reservation. We begin with a brief description of each of these categories (based on Luczak [LUCZ78] and then proceed with a discussion of models of access protocols.

CHANNEL ACCESS TECHNIQUES

Selection

Selection techniques are the oldest access protocols. Their use originated in controlling access to multipoint communications lines as well as computer buses. The essential feature of a selection technique is that each node on the network must receive permission to send its messages onto the network. Until it receives this permission, it must queue its messages.

Selection techniques may be centralized or decentralized (distributed). In the case of centralized control, a central channel controller grants this permission, while in the distributed case, control is distributed throughout the nodes. The three types of centralized selection techniques are *daisy-chaining, polling,* and *independent requests.*

Daisy-chaining is a technique employed on internal computer buses. The PDP11 Unibus and the IBM 370 I/O channel both employ this technique. The signal line is "daisy-chained" through the nodes for the purpose of selecting the bus master. When a signal reaches a node, it can become bus master by stopping the signal and then broadcasting its messages. In the event that its message queue is empty, it simply propagates the signal onto the next node.

Both the advantages and the disadvantages of daisy-chaining arise from its simplicity. On the positive side, it is an easy technique to implement. On the negative side,

1. it automatically imposes a fixed-priority structure,
2. its select pulses occasion time delays, and
3. it is vulnerable to failures in the grant line as well as the nodal interface.

In polling, a node is selected by being addressed. All nodes are informed of the next node to be selected. As in daisy-chaining, a node has the option of becoming bus master or of refusing control of the line if that node does not wish to send any messages. The central controller may question each node in turn, or—in the event of a prioritized nodal structure—it may question them in a sequence determined by the priorities. This form of polling is often called roll-call polling.

Polling may be implemented on any serial channel, and therefore (unlike daisy-chaining) requires no special grant lines.

In the independent-requests access technique, each node requests control of the bus from the central controller. The controller then ranks the requests according to their priorities, and selects the nodes accordingly. This system may be implemented in a variety of ways. Separate requests as well as separate select lines are one possibility on a parallel bus. Various time-multiplexing techniques may be employed on a serial channel for both requests and selections.

On a parallel bus, the independent requests method provides an efficient approach to the use of dynamic priority schemes. The major penalty to be paid, however, is the large number of control lines required for such a system.

Decentralized selection techniques that correspond to the three forms of centralized control have also been implemented (e.g., decentralized polling). A description of these techniques is presented by Luczak [LUCZ78].

Random Access

Random access techniques are characterized by a lack of strict ordering of the nodes contending for access to the channel. In a random access technique, a node is free to broadcast its messages at a time determined by the node without being absolutely certain that no other node is simultaneously attempting to broadcast. In fact, in the original implementation of a random access protocol (the ALOHA system at the University of Hawaii), a node was free to broadcast whenever it had messages to send.

There is a great deal of interest in random access techniques for bus networks, primarily because of the bursty nature of computer traffic [FUCH70]. A random access protocol provides the entire bandwidth of the channel to a user, *once he gains access to the channel*. Thus, if a small population contends for a channel at any given instant, a user with a message to transmit is guaranteed access to the full bandwidth of the channel after a brief waiting period.

The price to be paid for random access is that messages may collide in transit. Collisions generally result in messages being rendered unintelligible. Hence, techniques have been developed for limiting access to the bus (thereby reducing the number of collisions that may occur) as well as for providing a retransmission sequence for messages that have collided (collision resolution).

This section continues with a description of random access techniques because of their importance and the great activity in this area. In presenting a discussion of random access techniques, we rely on Luczak's taxonomy of random access protocols. Luczak's tree diagram of access control methods is presented in Fig. 41. Examples of systems employing those access methods are indicated on the diagram.

At the second level, random access techniques are either slotted or unslotted. In slotted techniques, all nodes are synchronized to a master clock. Time is subdivided into a collection of equal intervals. When a node has a message to send, it first subdivides the messages into packets of equal length (corresponding to the length of the time interval), which are then broadcast into the slots. If an acknowledgment for the packet is not received after some fixed period of time,* then the packet is assumed to have been destroyed by a collision, and is rebroadcast onto the network. The ALOHA technique is the epitome of random access techniques—if a node has a message to transmit, it simply does so. To improve its efficiency, a slotted version was developed (referred to as slotted ALOHA).

Unslotted techniques permit the transmission of variable-length messages. As indicated in Fig. 41, there are two major categories of unslotted messages:

1. pure ALOHA or "deaf" transmission, and
2. carrier sense multiple access (CSMA) techniques.

In CSMA, the sending node listens to the channel (sense carrier) before (and possibly during) message transmission. If carrier is sensed, the transmission is postponed for some period of time. If carrier is not sensed, the node does not have a guarantee that its message will arrive safely. The message is still vulnerable for a fraction of the time required to transmit it onto the network. This fraction of time corresponds to one propagation delay. Since the propagation delay

---

*In a dual, unidirectional channel, an acknowledgment may be obtained by listening to the receiving channel. The signal should arrive after one propagation delay, thereby providing an automatic acknowledgment.

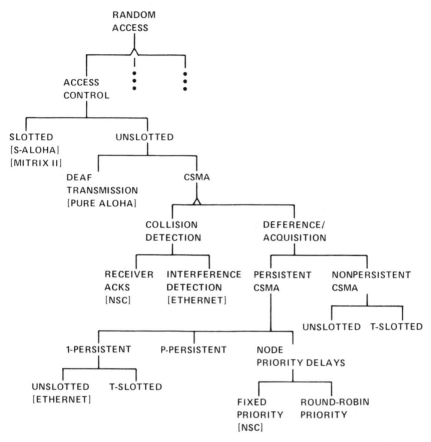

Fig. 41. Random access techniques—access control; NSC, Network Systems Corp. Hyper-channel. (Based on Luczak [LUCZ78].)

on a local bus network is considerably less than the transmission delay, the period of message vulnerability is brief. The objective of all CSMA protocols is to minimize the number of collisions that occur during this time window.

There are two fundamental facets of a CSMA protocol—the collision-detection method employed and the channel access technique employed (referred to as the deference/acquisition technique in Fig. 41). Collision detection, as the name implies, is the means by which a node discovers that a message which it has sent has collided with another message. This can be accomplished by an acknowledgment broadcast by the receiving node (discussed previously) or by the simple approach of having the node listen to the channel for a short time subsequent to message propagation.* This idea forms the basis of the listen-

---

*Twice the message propagation delay suffices to discover a collision.

while-talk protocol implemented on the MITRE bus system. When two or more users detect a collision, they immediately stop transmitting their message.

In order to understand how a node acquires control of the channel, one must know what the node does when the channel is sensed busy (deference), as well as what it does when the channel is sensed idle (acquisition). We shall discuss these access techniques first. The two major access techniques in a CSMA protocol are *persistent and nonpersistent CSMA*. A nonpersistent CSMA protocol may be described as follows.

A "ready" node (i.e., one with a message to transmit) senses the channel. Then:

1. If the channel is idle, the ready node transmits its message.
2. If the channel is busy, it reschedules the message according to its collision resolution algorithm. (Typically, it picks a value out of a retransmission delay distribution.) It then repeats step 1 after the expiration of this delay.

The node is called nonpersistent because it does not continue to sense the channel after it has determined the channel is busy.

A defect of this approach is that several nodes that have already been involved in a collision might not broadcast on an idle channel if their retransmission delays have not expired. This defect led to the introduction of persistent protocols, in which a node continues to sense a busy channel until it becomes idle, and then broadcasts its message. The obvious defect of such a protocol is that if two nodes have messages ready to transmit when the channel is busy, the messages will certainly collide when the channel becomes idle. To remedy this defect, a category of protocols with random transmission delays was introduced. Such protocols are called p-persistent protocols.

In a p-persistent protocol, time is divided into slots of length equal to the maximum propagation delay.

1. If a channel is sensed idle, it broadcasts a packet with probability p, and delays one slot with probability $(1 - p)$, at which point it again senses the channel. If the channel is sensed busy, it waits until the channel is sensed idle, and repeats this step.
2. If the channel is again sensed idle in the next slot, it repeats step 1.

Thus, the p-persistent protocols represent an attempt to use an idle channel as soon as possible and at the same time decrease the number of collisions such immediate use might cause.

The probability of transmission (p) is a small number—typical values are 0.03 and 0.1.

The last category of persistent CSMA protocols is built around the concept of nodal priorities. Here the idea is to define delay times based on priorities assigned to the nodes. If the channel is sensed idle, a ready node broadcasts its message. If it is sensed busy, the nodes delay an amount of time determined by their priorities. A sequence of delays $d_1, d_2, \ldots, d_N$ (N is the number of nodes) determines the order of broadcast.

For a more detailed account of these protocols (as well as other bus techniques), consult Luczak [LUCZ78].

Reservation

The third major technique for controlling channel access is known as reservation. In reservation techniques, a node transmits a message (or packet) in a slot that has been reserved for its use. In most reservation techniques, time is "slotted," giving rise to a packet-switching environment. Luczak's summary of the various reservation techniques according to their characteristics is shown in Fig. 42. Examples of systems employing these access methods are shown. As indicated in the figure, the major distinction between reservation techniques is whether they are *static* or *dynamic*.

Time-division multiple access (TDMA) is a well-known static approach. In TDMA, each node is assigned a fixed number of slots per frame. The slots may be assigned to each node according to its requirements. If the nodal requirements are known in advance, this results in high-channel utilization. On the other hand, if the data rates are bursty, channel utilization decreases. Digitized audio and video signals are applications which benefit from such a static approach.

In dynamic control, slots are assigned on a demand basis. The two fundamental divisions here are centralized and distributed control. Under centralized control, the nodes make their requests of a central controller, which in turn determines the appropriate number of slots for each node. Distributed control reservation schemes have largely been proposed for satellite systems, as the large propagation times involved in such systems would force a centralized controller to make its decisions on old information.

The two types of centralized control noted in Fig. 42 are connection-based control, in which a node requests transmission capability over a period of time, and message-based control, in which the node makes a reservation for each message. The original MITRE bus system, MITRIX 1, used a connection-based control system.

Distributed control techniques may be divided into explicit and implicit techniques, depending on whether a special message is or is not required to request slots. Explicit reservations are made via minislots preceding the message slots (reservation subframes) or are made within the message slot (piggyback).

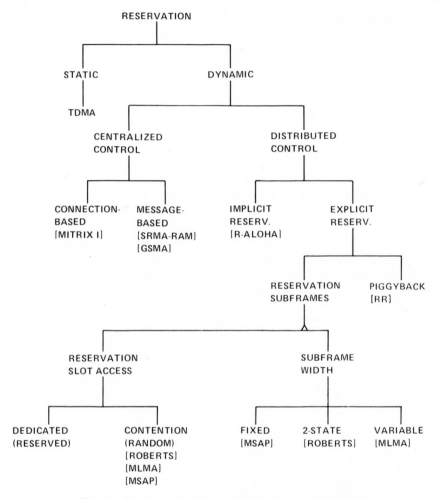

Fig. 42. Reservation techniques. (Based on Luczak [LUCZ78].)

When reservation subframes are employed, one must decide which access technique to employ as well as the width of the subframe.

REVIEW OF CLASSIC MODELS

We present a brief summary of the "classic" modeling work done on bus systems (up to 1976). A more detailed presentation of this work is given by Kleinrock [KLEI76].

ALOHA

The original work on the modeling of broadcast communications systems had its impetus in developing access schemes for satellite communications channels. The ALOHA system at the University of Hawaii inspired the first random access protocol [ABRA73]. Under the ALOHA protocol. if a packet requires P sec to be transmitted, it will be vulnerable during a period of $2P$ sec, as shown in Fig. 43.

Our model of the ALOHA system will be an infinite population model. Total offered traffic is G packets per transmission period P, where it is understood that each user contributes an infinitesimal amount to the total traffic G. Letting S denote the throughput (number of successful transmissions per P sec), Roberts [ROBE73] demonstrated that S and G are related by the equation

$$S = Ge^{-2G},$$

where e^{-2G} is the probability of successfully transmitting a packet. We may see this as follows:

the probability of successfully transmitting one packet
= the probability that no other packets arrive during the vulnerable period $2P$; assuming Poisson traffic, this probability
$= e^{-2G}$.

If we constrain packets to be transmitted only every P sec, then we *halve* the vulnerable period (P sec instead of $2P$ sec), and obtain corresponding improvement in the throughput; that is,

$$S = Ge^{-G}.$$

This relationship is demonstrated in Fig. 44. Note that the peak throughput for pure ALOHA is $1/2e$ (0.18), while for slotted ALOHA we have a peak throughput of $1/e$ (0.36).

Lam has developed a time delay versus throughput model for slotted ALOHA. His work originally focused on satellite channels but is equally applicable to local broadcast networks [LAM74].

As in the preceding throughput analysis, we assume an infinite population

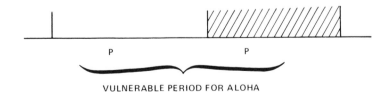

VULNERABLE PERIOD FOR ALOHA

Fig. 43.   Vulnerable period for ALOHA.

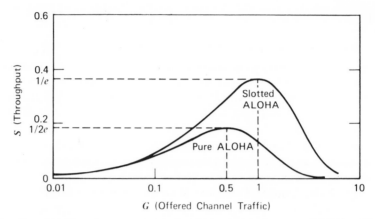

Fig. 44. Throughput for pure and slotted ALOHA. (From Kleinrock [KLEI76, p. 365]. Reprinted by permission.)

model. We define the average time delay T to be the average time (in slots) until a packet is successfully received. In our model, a user will broadcast a packet onto the network. If his packet is destroyed, the user will rebroadcast the packet randomly during one of the next K slots (with probability $1/K$ for each slot). This staggering of the packet transmission avoids the situation in which two users collide in a slot and then immediately rebroadcast their packets, thereby ensuring a second collision (and, in fact, an infinite number of collisions).

It is clear then that the effect of the number of retransmission slots K must be incorporated into a formula for the average packet time delay. Lam's equation for the average time delay T is therefore

$$T = 1 + E \left(\frac{K + 1}{2} \right),$$

where E is the average number of retransmission attempts per packet.

Lam then developed an expression for E in terms of the system parameters K and G as well as the throughput S. This expression is

$$E = \frac{1 - q}{q_t},$$

where

$$q = \left(e^{-G/K} + \frac{G}{K} e^{-G} \right)^K e^{-S},$$

and

$$q_t = \left(\frac{e^{-G/K} - e^{-G}}{1 - e^{-G}} \right) \left(e^{-G/K} + \frac{G}{K} e^{-G} \right)^{K-1} e^{-S}.$$

Lam also developed an expression for the throughput S in terms of q, q_t, and G, given by

$$S = G\frac{q_t}{1 + q_t - q} \, .$$

For a derivation of these equations, the reader may wish to consult Lam [LAM74] or Schwartz [SCHW77].

Ideally one would wish to solve the equations for q, q_t, and S simultaneously in terms of the system parameters G and K, thereby obtaining an expression for the time delay T in terms of these parameters. This is a difficult task; hence, one must settle for numerical results.

The fundamental relationships among the four parameters T, S, K, and G are depicted in Fig. 45. The dashed lines correspond to constant G (offered load) contours. We note that the effect of increasing the number of packet re-transmission slots K is to increase throughput.

Once K is increased beyond 15, the increase in throughput is infinitesimal, while the time delay increases precipitously. Fixing K at any value, we note that the maximum throughput occurs at $G = 1.0$. An attempt to increase the channel load beyond this point rapidly drives up the time delay.

Figure 46 shows an alternative way of viewing this trade-off. This diagram is interesting because of its constant time-delay contours. One can increase the throughput while maintaining a constant time delay by decreasing K—provided, again, one does not increase K too much. An optimal contour relating these three parameters is displayed in the figure.

Stability Considerations

As may be observed from Fig. 45 (which portrays the equilibrium time delay versus throughput trade-off) one can find two possible time delays for given values of S and K. The smaller of the two is called the channel operating point, because it represents the ideal point at which to operate the channel in light of the time delay versus throughput results. The existence of two time delays for a single throughput suggests that the channel equilibrium assumption is not valid. To investigate this possibility, Kleinrock and Lam simulated an infinite population model. The simulations revealed the following behavior: Beginning with an empty system, the channel proceeds into equilibrium at the channel operating point for a finite time period, after which stochastic variations in the traffic arrival pattern increase the traffic load, resulting in decreased through-put and higher packet delays. This pattern repeats itself until the channel even-tually drifts into saturation, with the throughput going to zero.

These observations prompted the realization that the fundamental trade-off

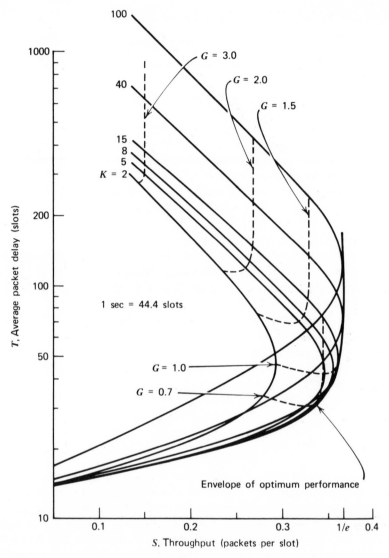

Fig. 45. Delay–throughput trade-off. (From Kleinrock [KLEI76, p. 374]. Reprinted by permission.)

for slotted ALOHA is not time delay versus throughput, but instead, *time delay versus throughput versus stability* [LAM74 or KLEI76].

In developing a model for this trade-off, Kleinrock and Lam employed a linear feedback model. The channel state for this model is portrayed as dependent on two random variables:

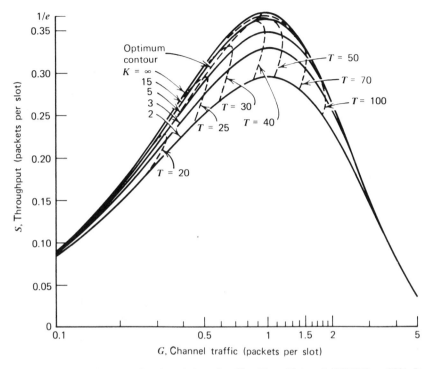

Fig. 46. Throughput as a function of channel traffic. (From Kleinrock [KLEI76, p. 373]. Reprinted by permission.)

1. $N(t)$, representing the total number of busy terminals at time t; and
2. $S(t)$, the combined input packet rate at time t.

If $N(t) = n$, then $S(t) = (M - n)\sigma$, where M is the number of active terminals (having a packet to transmit) and σ is the probability that a given terminal will transmit a packet in a slot.

Using this model, Kleinrock and Lam developed an equation representing the throughput in terms of the parameters N and S. This relationship is illustrated in Fig. 47. It is important to note that the throughput S_{out} represented in this figure is *not* the same as the equilibrium throughput S. In this context S is the amount of new channel traffic, that is, the input. The shaded region represents a safe region, in which the throughput (S_{out}) exceeds the input, whereas the unshaded region represents the situation in which the capacity of the system is exceeded.

With this in mind, two categories of channels can be defined:

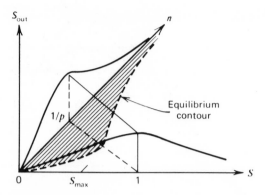

Fig. 47. Channel throughput rate as a function of load and backlog. (From Kleinrock [KLEI76, p. 378]. Reprinted by permission.)

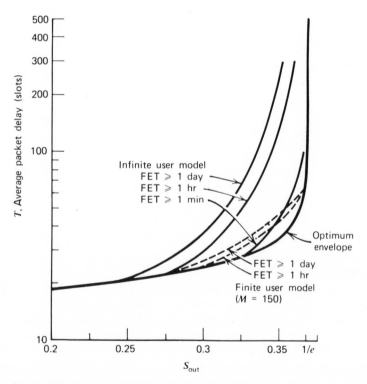

Fig. 48. Stability–throughput–delay trade-off. (From Kleinrock [KLEI76, p. 383]. Reprinted by permission.)

1. Stable channels, in which equilibrium results always obtain. A finite population of users falls into this category, *provided that a sufficiently high value for K is chosen*. Choosing higher values for K results, of course, in higher time delays.

2. Unstable channels, in which the equilibrium conditions holds only for a finite period of time, after which the channel drifts into an overload situation. As already indicated, the infinite population model is unstable.

To provide a numerical measure of instability, Lam defined the first exit time (FET) as the time required to exit into the unsafe region, starting from a zero backlog. With this definition in mind, it is possible to represent the fundamental time delay versus throughput versus stability trade-off by Fig. 48.

The lower solid curve represents the optimal envelope obtained from the time delay versus throughput trade-off. The upper three solid curves represent an infinite population model corresponding to three different FET values of ≥ 1 day, ≥ 1 hour, and ≥ 1 minute, while the two dashed curves represent a finite population model ($M = 150$) with FET values of ≥ 1 day and ≥ 1 hour.

As indicated by these curves, a better time delay versus throughput performance is achieved at the expense of more frequent channel overloads. If occasional channel overloads are acceptable, then the channel need only be shut down and restarted when such a situation occurs. If this solution is not acceptable, Lam has developed dynamic control procedures to ensure system stability [LAM74].

MITRE Cable Bus System

The slotted ALOHA protocol has been implemented on a cable bus system (referred to as MITRIX 2) developed at The MITRE Corporation, Bedford, Massachusetts. Each user in the system is attached via an interface unit to two unidirectional cables. One of the cables is employed to carry data from the users to a repeater located at the head-end of the system. An outgoing cable carries the messages from the repeater back to the users. The transmission rate is 7.373 Mbit/sec. Figure 49 is a diagram of this system.

MITRIX 2 is a dual-mode system, built to accommodate both high and low duty-cycle users. Synchronous, high duty-cycle users have dedicated slots assigned to them while the bursty, low duty-cycle users are assigned a set of slots which they compete for in ALOHA fashion. Control procedures based upon Lam and Kleinrock's work are employed on the ALOHA subchannel. A description of the system is presented by Meisner *et al.* [MEIS77].

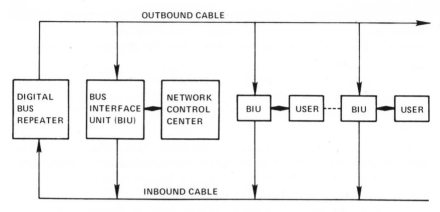

Fig. 49. MITRIX bus structure. (Based on Meisner *et al.* [MEIS77, p. 5–14].)

Modeling of CSMA Protocols

As mentioned previously, with CSMA protocols, the users listen to the channel and wait until it is idle before broadcasting their messages onto the network. The impetus for a protocol of this nature lies in the fact that the propagation delay in a local network is much smaller than the transmission delay, and one may therefore obtain up-to-date information about the state of the channel. For example, if a 1000-bit packet travels over a 10-mile, 100 kbit/sec channel, one obtains a transmission time of 10 msec and a maximum propagation delay of 0.054 msec. This situation is a great contrast to a satellite communications channel, in which the propagation delays are of the order of 0.25 sec. Listening to a satellite channel provides information about the past. In fact, it provides such ancient history that attempting to employ a CSMA protocol on such a channel would be deleterious, since decisions would be made on out-of-date information.

An extension of the CSMA protocols has been developed in which the users continue to monitor the transmission line during their message transmission and immediately cease transmission on detecting a collision. This protocol is called the listen-while-talk (LWT) protocol.

The two major categories of persistent and nonpersistent CSMA protocols have already been discussed in some detail; we shall now simply restate the algorithms for nonpersistent, 1-persistent, and p-persistent CSMA in order to refresh the reader's memory.

In a nonpersistent protocol, a ready node (one having a message to transmit) first senses the channel. Then

1. If the channel is sensed idle, it transmits its message.

2. If the channel is sensed busy, it reschedules the message for some later time, picking a delay time from a retransmission distribution. It then repeats step 1.

In a 1-persistent protocol, a node with a message to transmit continues to sense a busy channel until the channel becomes idle. Then the node transmits a message.

In a p-persistent protocol, time is slotted into slots of length equal to the maximum propagation delay. A ready node senses the channel, and:

1. If it is idle, the node broadcasts the packet with probability p, and delays one slot with probability $(1 - p)$, at which point it again senses the channel. If busy, it continues to sense the channel, and repeats this step.

2. If the node is delayed one slot, it then repeats step 1.

Our first category of models is aimed at developing a throughput (denoted by S) versus offered load (denoted by G) trade-off. In doing so, we develop throughput equations for S in terms of G and other system parameters. These equations were developed by Tobagi [TOBA74].

In developing these equations, Tobagi assumes an infinite population model generating an average of G packets per P sec. Other assumptions are the following:

1. Each packet is of constant length.

2. Each terminal has *exactly* one packet awaiting transmission.

3. The acknowledgment channel is separate from the data channel.

4. All source-destination pairs are assumed to have the same one-way delay. This delay is normalized with respect to transmission time on the channel.

A complete list of the assumptions of the model are given by Tobagi [TOBA74].

The propagation delay is denoted by a and is equal to $(d/2)/P$, where $d/2$ is the one-way propagation delay and P is the packet transmission time.

We develop an equation for S in terms of G for nonpersistent CSMA to give an indication of the modeling approach employed in developing the throughput equations for the CSMA protocols. The interested reader may consult Tobagi [TOBA74] for the remaining equations as well as their derivations. LaBarre [LABA78] points out an error in Tobagi's throughput equation for nonpersistent CSMA and corrects it. Hence we present LaBarre's derivation of this equation.

In analyzing nonpersistent CSMA, LaBarre notes that the activity on the channel may be divided into busy and idle periods. The combination of one busy and one idle period is the channel cycle.

The following equation from renewal theory enables LaBarre to calculate the throughput:

$$S = \frac{\overline{U}}{\overline{B} + \overline{I}},$$

where $\overline{U}$ is the average time during the cycle in which the channel is used successfully, $\overline{B}$ the average length of the busy period, and $\overline{I}$ the average length of the idle period.

To calculate these quantities, we refer to Fig. 50. In this figure, a packet is pictured arriving at time 0. It is subject to a number of collisions during the first busy period, resulting in an unsuccessful transmission period. The last arrival occupies the channel for 1 (normalized) transmission delay, and clears the channel a (normalized) time units later.

The busy period is followed by an idle period (mean length $1/G$) and is immediately followed by another busy period. In this busy period, the transmission is successful; hence, the length of this period is $1 + a$.

We start with $\overline{U}$. The fraction of time during a cycle in which the channel is used without interference is equal to the probability that no packets arrive during the first seconds of the transmission of the packet. Since we assume an average arrival rate of G packets per P sec, the Poisson distribution yields e^{-aG}. (Recall that $a = d/2P$.)

To calculate $\overline{I}$, we note that the interarrival time must be $1/G$ (Poisson process).

To calculate $\overline{B}$, we first define the random variable Y in the interval $(0, a)$ to be the instant at which the last packet that collides with our first arrival appears. If there are no collisions with our first arrival (i.e., a successful transmission), then the length of the busy period is $1 + a$. The average length of

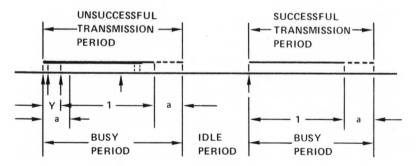

Fig. 50. Nonpersistent CSMA: busy and idle periods. (Based on Tobagi [TOBA74, p. 55].)

the unsuccessful busy period is equal to $1 + \overline{Y} + a$, where $\overline{Y}$ is the mean value of Y. We now compute $\overline{Y}$.

The distribution function $F_Y(y)$ is

$$F_Y(y) = P(\text{no arrivals occur in an interval of length } a - y)$$
$$= e^{-(a-y)G}.$$

The mean of this distribution, $\overline{Y}$, is

$$\overline{Y} = a - \frac{1}{G}(1 - e^{-aG}).$$

$\overline{B}$ may now be calculated as

$$\overline{B} = P(\text{successful transmission})(1 + a) + P(\text{unsuccessful transmission})(1 + \overline{Y} + a)$$
$$= e^{-aG}(1 + a) + (1 - e^{-aG})(1 + \overline{Y} + a)$$
$$= 1 + (1 - e^{-aG})\overline{Y} + a.$$

Substituting values for $\overline{I}$, $\overline{B}$, and $\overline{U}$ in our throughput equation, we obtain for S,

$$S = \frac{Ge^{-aG}}{G(1 + 2a - ae^{-aG}) - (1 - e^{-aG})^2 + 1}.$$

Tobagi obtains throughput equations for the other CSMA protocols he defines [TOBA74] (1-persistent and p-persistent CSMA), as well as slotted versions of these protocols. The interested reader should consult Tobagi [TOBA74] or Kleinrock and Tobagi [KLEI75c].

LaBarre [LABA78] develops a throughput equation for the nonpersistent listen-while-talk (LWT) CSMA protocol. His equation is

$$S = \frac{Ge^{-aG}}{G(e^{-aG} + a) + (1 + aG)(1 - e^{-aG})^2 + 1}.$$

LaBarre's graph comparing the nonpersistent LWT protocol with the other CSMA protocols is shown in Fig. 51.

LaBarre points out that the nonpersistent LWT protocol offers a 10–30% improvement in maximum throughput for propagation delays a ranging between 0.01 and 0.05. Tobagi presents a similar diagram (Fig. 52) which includes p-persistent CSMA as well as slotted nonpersistent CSMA.

As can be seen by comparing Figs. 51 and 52, nonpersistent LWT also has the edge in throughput performance over the p-persistent CSMA protocols.

It is important to note the effect of parameter a on the throughput of these protocols. Increasing the value of a does not affect the ALOHA protocols, but it does significantly affect the CSMA protocols. The reason (pictured in Fig. 53) is that the large values of a correspond to older information. Hence, decisions on

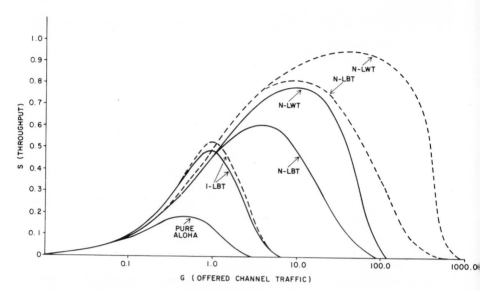

Fig. 51. Throughput versus offered load—analytic results: ---, $a = 0.01$; —, $a = 0.05$; N-LBT, nonpersistent LBT; N-LWT, nonpersistent LWT; 1-LBT, 1-persistent LBT.

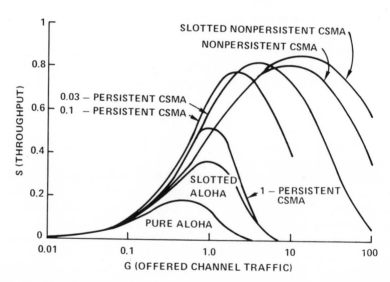

Fig. 52. Throughput versus channel traffic; $a = 0.01$. (Based on Kleinrock [KLEI76, p. 400]. Used by permission.)

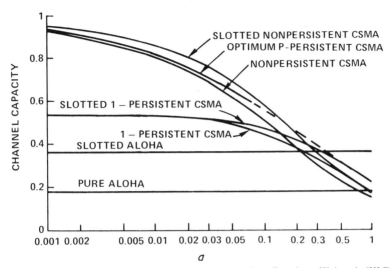

Fig. 53. Effect of propagation delay on channel capacity. (Based on Kleinrock [KLEI76, p. 400]. Used by permission.)

whether or not to broadcast a packet will be made on incorrect information. (Never trust second-hand information—it degrades performance.)

We now turn to time delay versus throughput models. Making two additional assumptions (to be discussed), Tobagi develops equations for the time delay in terms of the system throughput, the offered load, and other system parameters. We present his equation for nonpersistent CSMA, and refer the reader to Tobagi [TOBA74] or Kleinrock and Tobagi [KLEI75c] for the equations of the remaining CSMA protocols, as well as their derivations.

In order to derive the equation, Tobagi assumes that

1. The average retransmission delay $\bar{X}$ is large when compared to the transmission time T.

2. The interarrival times of the point process (defined by the start times of all the packets plus transmissions) are independent and exponentially distributed.

3. The acknowledgment packets are correctly received with probability 1. (One might create a separate channel for acknowledgments.)

4. The processing time required to perform the sumcheck and to generate the acknowledgment packet is negligible.

Let W be the (normalized) time delay for an acknowledgment packet, and let ξ be the normalized mean retransmission delay. Then the expected time delay D must be the sum of

1. The transmission time, given by $1 + a$.

2. The expected delay due to deferring (i.e., sensing a busy channel). The expression for this mean delay is $[(G - H)/S]\xi$, where H is the amount of traffic the channel attempts to transmit.

3. The expected retransmission time, given by $[(H - S)/S](1 + 2a + W + \xi)$. The second term in parentheses represents the sum of a transmission, an acknowledgment, and a retransmission delay.

Hence, the expression obtained for D is

$$D = 1 + a + \left(\frac{H - S}{S} \right)(1 + 2a + W + \xi) + \left(\frac{G - H}{S} \right) \xi.$$

H may be computed as

$$H = G(1 - P_b),$$

where P_b is the probability of being blocked. An expression for $1 - P_b$ is given by

$$\frac{1 + aG}{1 + G(1 + a + \bar{Y})}.$$

It is possible (although difficult) to obtain optimal values of ξ such that D is minimized for a given value of S. This may be seen from the fact that G/S is a decreasing function of ξ.

A simulation was performed which presents the basic time delay versus throughput trade-off but with assumptions 1 and 2 dropped. The result of this simulation is presented in Fig. 54 (for $a = 0.01$).

The best performance is obtained from the optimal p-persistent protocol.

LaBarre [LABA78] develops an equation for nonpersistent LWT. He then uses the analytic delay results to perform the same trade-off. However, he replaces the p-persistent protocol with the LWT protocol, and concludes that the LWT protocol is best. His graph is reproduced in Fig. 55.

The 1-persistent LWT protocol has been implemented on the MITRE bus system.

Simulation studies of the system were performed in which the 1-persistent LBT protocol was compared to the 1-persistent LWT protocol [LABA78]. The results indicated that the LWT protocol has a maximum throughput of more than twice that of the LBT protocol, and a time delay of less than half of the LBT protocol for corresponding throughput values.

Tobagi and LaBarre point out that the dynamic control procedures developed by Lam for use in a slotted ALOHA channel are applicable to the CSMA and LWT protocols [TOBA74, LABA78].

The LWT protocol has been implemented in three systems:

1. *Ethernet*   Developed at the Xerox Corporation, Ethernet connects up to 256 communicating computers at 3 Mbit/sec over 1 km of coaxial cable, utilizing off-the-shelf CATV taps and connectors. A description of Ethernet is presented by Metcalfe and Boggs [METC76].

2. *Fibernet*   Developed at the Xerox Corporation, Fibernet makes use of optical fibers to connect up to 19 stations at 150 Mbit/sec through $\frac{1}{2}$ km of optical fiber. A description of Fibernet is given by Rawson and Metcalfe [RAWS78].

3. *MITRE Cable Bus*   A brief description of this sytem is given in the section of this chapter on the ALOHA protocol.

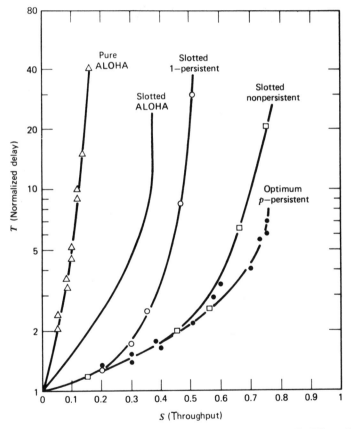

Fig. 54. Time delay versus throughput, $a = 0.01$. (From Kleinrock [KLEI76, p. 401]. Reprinted by permission.)

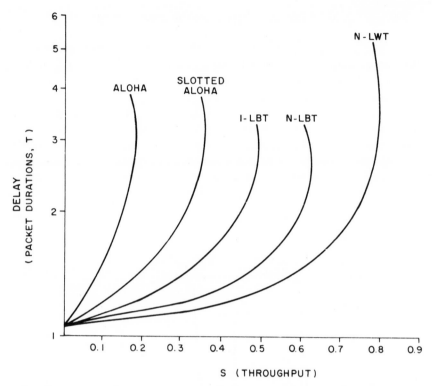

Fig. 55. Analytic delay results: $\alpha = 0.05$; $\bar{\xi} = 0.12$; N-LBT, nonpersistent LBT; N-LWT, nonpersistent LWT; 1-LBT, 1-persistent LBT. (Based on LaBarre [LABA78].)

HYPERCHANNEL

Network Systems Corporation of St. Paul, Minnesota, has designed a local network called Hyperchannel which employs a prioritized CSMA access protocol.

This network was developed to alleviate the bottleneck caused in many computer centers when one large computer was assigned the task of controlling numerous storage devices for the remaining processors on the site.

In Hyperchannel, each port connects through an adapter (which performs functions such as buffering and flow control) to as many as four coaxial cables, each having a line speed of 50 Mbit/sec. Each coaxial cable can accommodate 16 hosts up to 1000 ft apart. Figure 56 is a diagram of this system. For more details, the reader should consult Thornton *et al.* [THOR75].

The access protocol employed by Hyperchannel is now described.

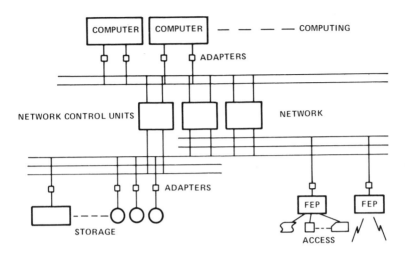

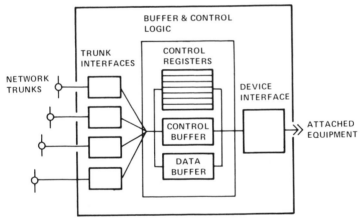

Fig. 56. Hyperchannel architecture. In lower diagram, one or more serial trunks are connected to conventional parallel devices through an adapter that converts the data, checks for errors, preprocesses the message, and communicates with peripheral controllers or processor channels. (Based on Thornton *et al.* [THOR75, p. 81].)

For the *i*th CIU* we have

Loop: *Wait until* message to transmit;

Loop 2:   *If* carrier absent *then* [transmit message; *if* collision *then*
[wait D_i time units; *goto* Loop 2]
else goto Loop]

*In this description, the adapter is referred to as a CIU (communications interface unit).

else [*wait until* carrier absent; wait D_i
time units; *goto* Loop]

goto Loop:

$$0 < D_1 < D_2 < \cdots < D_n$$

$$D_{i+1} - D_i > d$$

with d representing maximum length path propagation delay [FRAN77, p. 3].

Extensive modeling of Hyperchannel has been performed. A queueing model was developed by Franta and Bilodeau [FRAN77], and a number of simulation studies have been conducted at the Lawrence Livermore Laboratory of the University of California [DONN78a,b, NESS78]. We shall discuss both the queueing model and the simulation studies and compare their predictions about the system.

Franta and Bilodeau were interested in obtaining an expression for the steady-state time delay suffered by the ith CIU in the system, as well as the steady-state throughput. They based their approach on Tobagi's analyses of both persistent and nonpersistent CSMA protocols [TOBA74]. This approach is based on renewal theory [COX62]. A cycle on the channel is pictured as consisting of alternating idle and busy periods.

The steady-state throughput is given by

$$S = \sum_i P[X = i] \frac{U(i)}{I(i) + B(i)} .$$

In this formula:

1. i is the state of the system. This state is defined as a vector with N components, where N is the number of CIUs sharing the channel. The state of each of the N CIUs sharing the channel may be represented by a 0, a 1, or a 2—indicating that the CIU is idle, that it is blocked,* or that it has started the current busy period.

2. $U(i)$ is the period of time during a cycle that the channel is conflict-free.
3. $B(i)$ is the length of the busy period.
4. $I(i)$ is the length of the idle period.

Each of these steady-state expressions is derived by obtaining an expression for the appropriate behaviors, and then passing to the limit.

To obtain the transient expressions for the nth cycle, one must evaluate two categories of expressions:

---

*That is, has a message to transmit, but cannot do so because the channel is busy. A 1 is also used to denote that a transmission was initiated by the CIU, but that it will be unsuccessful.

1. transition probabilities of the form

$$P(X_{n,k} = j/X_{n,k-i} = i),$$

where $X_{n,k}$ represents the state of the system at the kth stage of the nth cycle;
2. time delays that occur between stages of the cycle.

The interested reader should consult Franta and Bilodeau [FRAN77] for the derivations of the quantities.

In order to derive the steady-state probabilities $P(X = i)$ that are used in the throughput equation, Franta and Bilodeau solve the classic equation for steady-state probabilities in a discrete-time Markov chain, that is,

$$\pi P = \pi.$$

Note that this involves solving 2^N equations in 2^N unknowns (N is the number of CIUs)—the authors have chosen the brute-force approach.

Next, the authors obtain an expression for the waiting time W_i of the ith CIU, defined to be the interval between the time when the CIU first makes the transition from the idle state to the busy state and the time when it makes the reverse transition.

W_i is calculated via the formula

$$W_i = \sum_{j=0}^{\infty} Q_i(j)R_i(j),$$

where j is the jth busy period, $Q_i(j)$ the probability that j successfully transmits in the jth busy period following becoming busy, and $R_i(j)$ the length of time between becoming busy and becoming idle.

After solving the preceding collection of equations, Franta and Bilodeau present a number of conclusions concerning the behavior of Hyperchannel. We begin with their discussion of *throughput versus offered load*. Figure 57 is based on an illustration from their paper.

As shown in Fig. 57, the relationship between throughput and offered load may be divided into three regions: In the first region the throughput increases with increasing load. The idle period on the channel decreases with increasing load, thus permitting the throughput to grow. (The utilization and the busy period remain relatively constant.)

In the second region, the idle period can no longer be decreased to accommodate increasing traffic, as the entire bandwidth of the channel has been used. An increased number of collisions results, and *throughput drops*.

Region three contains the apparent paradox of increased throughput at still higher loads. The explanation for this, however, lies in the priority access protocol of Hyperchannel. With increased loads, the CIU with the highest priority dominates the channel to the exclusion of the remaining CIUs.

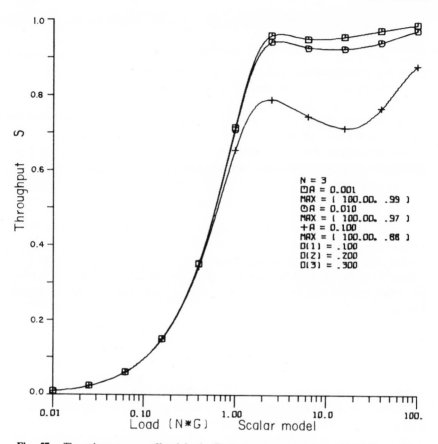

Fig. 57. Throughput versus offered load. (From Franta and Bilodeau [FRAN77, p. 44].)

TABLE 2

Traffic Load[a]

γ (sec)	Rate = $1/\gamma$	Load = adapters · rate
50,000E − 6	20	6E + 1
10,000E − 6	100	3E + 2
2000E − 6	500	1.5E + 3
400E − 6	2500	7.5E + 3
80E − 6	12,500	3.75E + 4
16E − 6	62,500	1.875E + 5
3.2E − 6	312,500	9.375E + 5
0.64E − 6	1,562,500	4.6875E + 6

[a] From Donnelley and Yeh [DONN78a, p. 129].

With respect to time delay/throughput trade-offs, Hyperchannel behaves in very much the same fashion as other random access protocols (slotted ALOHA, CSMA). That is, in order to efficiently operate Hyperchannel, one must search for the "knee" of the time delay/throughput curve. The highest throughputs once again correspond to the highest time delays. Consult Kleinrock [KLEI76] for a discussion of the throughput/time delay/stability trade-offs in random access protocols. A summary of these trade-offs was given earlier in this chapter.

Simulation studies conducted by Donnelley and Yeh at Lawrence Livermore Laboratories [DONN78a] are in close agreement with the results of Franta and Bilodeau. Several performance curves from these studies are reproduced in Table

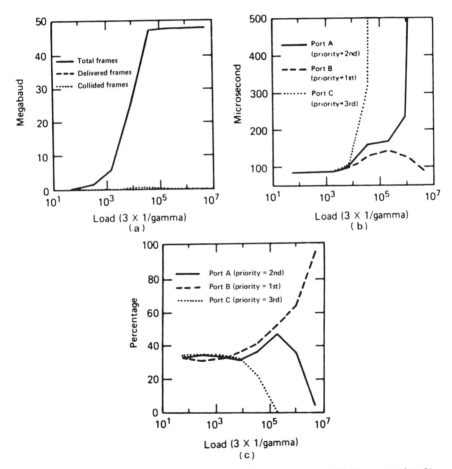

Fig. 58. Performance curves (semilog plot) for three-port network (4K-bit message length): (a) trunk throughput; (b) frame delay of each port, (c) throughput percentage of each port. (From Donnelley and Yeh [DONN78a, p. 129].)

2 and Fig. 58. The simulations were written in ASPOL, and run on a CDC 7600 computer.

More detailed simulations were conducted by Donnelley and Yeh on the higher level protocols governing flow control, recovery from errors, etc. (termed level 2 protocols by these authors). This work was continued by Donnelley and Yeh [DONN78b]. In both papers, the authors point out problems (and suggested solutions) in the level 2 protocol *which tended to negate the good performance of the level 1 protocol.* Their major conclusion is that it is vital to construct level 2 protocols in a prioritized CSMA system such as Hyperchannel in such a way as to complement the level 1 (trunk access) protocol [DONN78b]. Nessett [NESS78] introduces a second-level protocol to alleviate some of the problems uncovered by Donnelley and Yeh [DONN78a,b].

CONFLICT-FREE RESERVATION SCHEMES

All the access schemes discussed so far have been characterized by messages subject to collision during a certain period of time, with a resulting loss in throughput. Some schemes have recently been introduced in an attempt to eliminate collisions, and thereby improve system performance. The salient feature of these schemes is that they employ some form of message reservation scheme. Time is divided into frames, which are in turn subdivided into reservation slots and message slots.

As their name indicates, reservation slots are employed by each user to reserve space for the packets to be sent. (We assume that messages are divided into packets.) The message slots are used to accommodate the user-generated packets. Control of these systems may be either centralized or distributed. In a centralized environment, a scheduler allocates message slots on the basis of requests, while in a distributed system, some form of nodal prioritization is employed.

Other reservation schemes were proposed earlier than those to be discussed here. Roberts proposed a centralized system for satellite packet switching, in which reservation slots were contended for in ALOHA fashion [ROBE73]. A centralized FDMA reservation scheme referred to as split-channel multiple access (SRMA) was proposed by Tobagi for packet radio systems [TOBA74]. In SRMA, two channels are created: a contention channel for message reservation and a TDMA channel for message transfer. The schemes of both Roberts and Tobagi employ random access methods for message reservations, while the schemes discussed in this chapter employ other techniques to resolve contention for message reservation.

We discuss six schemes. The first, the minislotted alternating priorities

(MSAP) scheme, was developed by Kleinrock and Scholl at the University of California, Los Angeles. It is used in a decentralized environment and is especially effective for a small number of buffered users (≤ 50). A discussion of a family of protocols closely related to MSAP is also presented, as these protocols provided the motivation for the development of MSAP.

Two close relatives of MSAP, BRAM, and COLUMBIA are then described.

The MLMA (multilevel multiaccess) protocol, developed at IBM (Zurich) is discussed next. This will be seen to be related to the precursors of MSAP. MSAP is intended for use in a distributed environment.

The fifth protocol to be discussed will be the GSMA (global scheduling multiple access) protocol, developed at the IBM Watson Laboratories. GSMA employs a centralized scheduling mechanism.

The last protocol discussed is the DYN (dynamic reservation) protocol, suggested by Kleinrock. DYN is intended for use in a distributed environment, and is shown to be particularly effective in a heavy traffic environment.

MSAP

Our description of MSAP (minislotted alternating priorities) is based on work by Kleinrock and Scholl [KLEI77b].

MSAP grew out of an attempt to provide good delay performance for a small number of users (~ 20) at higher traffic loads. The CSMA protocols tend to provide good delay performance at low loads, but performance degrades at higher loads as a result of the increasing number of collisions. At the other extreme, one finds the fixed assignment systems of time-division multiple access (TDMA) and frequency-division multiple access (FDMA), which are suitable for high traffic levels but perform poorly at lower traffic levels.

A family of three protocols was developed that solved the preceding problem. Known as the alternating priorities (AP), round robin (RR), and random order (RO) protocols, they all use carrier sensing in the following manner.

Time is divided into frames, as indicated in Fig. 59. The minislots serve as reservation slots for the users. Each slot has a duration equal to the maximum propagation time (τ). The nodes are arranged in some sequence in which they may broadcast. If a node wishes to broadcast a packet, it simply sends a signal on the line. If it does not wish to broadcast, it remains silent. Hence, after a duration of τ sec, all the nodes are aware of the intentions of this particular node. In effect, the token (i.e., control of the line) is passed to the next user in the form of silence.*

---

*The scheme is clearly similar to hub polling, in which control passes from user to user [SCHW77], with the major exception that in this case the control is decentralized.

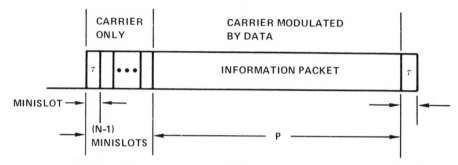

Fig. 59. Slot configuration in AP, RR, and RO. (Based on Kleinrock and Scholl [KLEI77b].)

As shown in Fig. 59, "space" is provided for one packet, followed by a slot of duration τ in which the packet clears the line.

Following Kleinrock and Scholl, the access scheme(s) may be formally characterized as follows:

The N users are ordered in each slot by the priority rule which characterizes the protocol. For all priority rules (and thus for all protocols), the N users are synchronized as follows in each slot:

1. If the highest priority user is ready, he need not sense the channel. He synchronizes the transmission of his packet as follows:

   a. At the beginning of the slot, he begins transmission of the carrier (with no modulation).
   b. $(N - 1)$ minislots later, he transmits his packet. Otherwise (if he is idle), he remains quiet until the end of the slot.

2. If the ith user in priority ($1 < i \leq N$) is ready, he senses the channel for $(i - 1)$ minislots.

   a. If no carrier is detected after $(i - 1)$ minislots, then at the beginning of the ith minislot, he transmits his carrier, and $(N - i)$ minislots later, he transmits his packet.
   b. Otherwise (idle user or carrier detected earlier), he waits for the next slot and the process is repeated (with a possibly different priority order).

Under all protocols, a slot is unused if and only if all users are idle. [KLEI77b].)

We also present Kleinrock and Scholl's descriptions of the priority structures which determine the order of transmission.

In the AP (alternating priorities) protocol, the N users are ordered according to a fixed sequence, $i = 1, \ldots, N$. Then:

1. Assign the slot to that user (say user i) who transmitted the last packet. If user i is ready, he transmits a packet in this slot. Otherwise (if there are no more packets from this user),
2. Assign the slot to the next user in sequence [i.e., user $(i \bmod N) + 1$].

    a. If this next user is ready, he transmits a packet in this slot, and in the following slot, operates as above.
    b. If this next user is idle, then repeat step 2 until either a ready user is found or the N users have been scanned. In this latter case (all users idle), the slot is unused and in the following slot, operates as above (the following slot is assigned to user i) [KLEI77b].

In the RR (round robin) protocol, users are assigned slots in cyclic fashion. As in TDMA, each user is preassigned one slot in a round robin (i.e., cyclic) fashion according to a given sequence of, say, $(1, 2, \ldots, N, 1, 2, \ldots)$.

1. If the user (say i) to whom the current slot is assigned is ready, he transmits a packet in this slot.
2. Otherwise (user i idle), assign the slot to the next user in sequence [i.e., user $(i \bmod N) + 1$].

    a. If this next user is ready, he transmits a packet in this slot.
    b. If this next user is idle, then repeat step 2 until either a ready user is found or the N users have been scanned. In this latter case (all users idle), the slot is unused.

3. No matter who uses the current slot (assigned to user i), the next slot is assigned to user $(i \bmod N) + 1$ [KLEI77b].

The RO (random order) protocol assigns priorities to the nodes randomly. Each user generates the same pseudorandom permutation of the digits $1, \ldots, N$ in order to determine which node has access to the next slot. A new user is chosen in this manner irrespective of who had use of the last slot.

Kleinrock and Scholl developed models to evaluate the channel capacity (maximum channel utilization) as well as the time delay/throughput trade-off. In addition, they compared the performance of these three protocols and the performance of other access methods (CSMA, TDMA, and polling). A brief summary of these models is followed by an evaluation of the three protocols.

Channel Capacity

Since $N\tau$ sec are lost each frame, the channel capacity C of all three schemes is

$$C = \frac{1}{1 + Na}.$$

From this equation, it follows that the capacity of these protocols will be large if either the number of users (N) is relatively small ($N = 10$) or if the value of a is small (0.001). If, for example, $a = 0.001$, then the capacity will be more than 90% for $N < 110$. More details are given by Kleinrock and Scholl [KLEI77b].

Packet Delay

All three schemes may be modeled as $M/D/1$* priority queueing systems with a rest period (corresponding to the reservation slot). The total packet input rate is denoted by λ packets/sec. It is shown [SCHO76] that for equal input rates $\lambda_i = \lambda/N$, $i = 1, \ldots, N$, the delay is given by

$$D_i = \frac{1}{2(1 - p)} + 1 \qquad \text{for} \quad i = 1, \ldots, N,$$

where $p = \lambda P (1 + Na)$ is the total normalized input rate (packets/slot). Hence the delay is independent of the protocol chosen in the equal input case.

Because the derivation of a delay formula in the case of unequal inputs is difficult, simulation was employed to obtain a better picture of the delay/throughput performance of the three protocols. The conclusion reached as a result of the simulation was that there was very little difference in the mean delay/throughput performance of the three protocols, and that there was only a small difference with respect to the variance.

The delay/throughput performance of the AP, RR, and RO protocols was compared to the performance of several other access schemes—CSMA, TDMA, and polling. The results are illustrated in Figs. 60 and 61. For a discussion of the models employed for the various access schemes in obtaining this graph, the reader is referred to Scholl [SCHO76] or Kleinrock and Scholl [KLEI77b].

Figure 60 illustrates that when there are a small number of users ($N = 10$) reasonably close together ($a = 0.01$), AP, RR, and RO perform better than CSMA in a high-traffic environment. For a larger numbr of users (50), the performance profile changes a great deal, as indicated by Fig. 61.

---

*$M/D/1$ indicates exponential arrivals/deterministic service/one server.

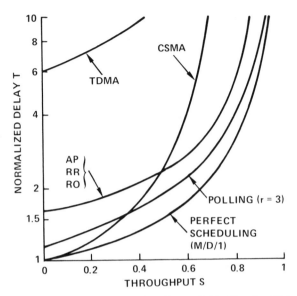

Fig. 60. Packet delay versus throughput ($N = 10$, $a = 0.01$). (Based on Kleinrock and Scholl [KLEI77b].)

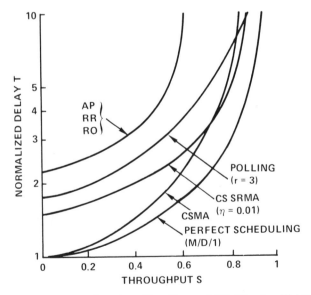

Fig. 61. Packet delay versus throughput ($N = 50$, $a = 0.01$). (Based on Kleinrock and Scholl [KLEI77b].)

In summarizing the results of their analyses, Kleinrock and Scholl make the following points:

1. For a small number of users, AP, RR, and RO provide a good channel capacity and a delay/throughput performance comparable to polling or CS SRMA (Fig. 61).
2. For sufficiently small a (e.g., $a = 0.001$), that is, when the users are sufficiently close together, the number of users can be reasonably large (50) without significant performance degradation. For larger a (e.g., $a = 0.01$), the performance degrades with the number of users, as indicated in Fig. 60.

It is clear from examination of the three protocols that a certain amount of channel time is consumed during the reservation portion of the frame. In order to reduce this time, the authors developed the MSAP protocol. The feature which distinguishes this protocol from AP, RO, and RR is that a user begins his transmission in a packet slot immediately after indicating that he has something to transmit in a reservation slot. The users are assigned a priority structure as with the preceding three protocols. The protocol obeys the AP structure (i.e., the last user to transmit retains control of the channel if his buffer is not empty). More formally, the protocol operates as follows. Assume that user i is the last user to transmit a packet.

By carrier sensing, at most one (mini-) slot later, all users detect the end of transmission of user i (absence of carrier); in particular, so does the next user in sequence [user $(i \bmod N) + 1$]. Then

a. either: User $(i \bmod N) + 1$ starts transmission of a packet; in this case, one slot after the beginning of his transmission, all others detect the carrier. They wait until the end of the transmission of this packet and then operate as above.
b. or: User $(i \bmod N) + 1$ is idle; in this case, one slot later, all other users do not detect the carrier; they know that it is the turn of the next user in sequence, that is, user $(i \bmod N) + 2$ and operate as above [KLEI77b].

The operation of the MSAP is illustrated in Fig. 62. In this example, two slots after user 3's transmission, user 1 gains control of the line as he detects that user 4 is idle. User 1 then transmits three packets, and is followed in turn by users 2 and 3. This example underlines the fact that the overhead for MSAP is significantly smaller than that for the AP, RR, or RO protocols. The only time lost on the channel under MSAP is one minislot due to switchover from one user to another. This is in contrast to the family of three protocols, in which N (number of users) minislots are lost at each packet transmission.

In order to obtain an equation for the time delay, we note the similarity of

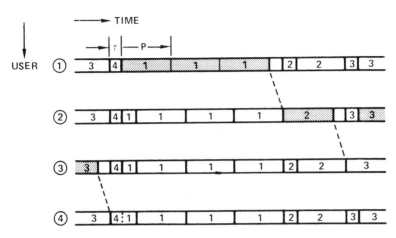

Fig. 62.   MSAP, four users. (Based on Kleinrock and Scholl [KLEI77b].)

MSAP to roll-call polling [SCHW77]. The difference between the two schemes is due to the large changeover time between two users in polling. In roll-call polling, this changeover time equals the polling message transmission time ($\geqslant 1$ slot) plus twice the propagation time between the user and the central station. Kamheim and Meister's results on roll-call polling may then be applied by simply setting the polling time equal to 1 slot in their equation [KONH74]. The expected packet delay T is then

$$T = 1 + \frac{S}{2(1 - S)} + \frac{a}{2}\left(1 - \frac{S}{N}\right)\left(1 + \frac{N}{1 - S}\right),$$

where S is the throughput, measured in packets generated per transmission time P.

By way of making a comparison between MSAP and other access methods, Kleinrock and Scholl provide a graph (Fig. 63). In this figure, it is assumed that there are 50 users ($N = 50$) and that $a = 0.01$.

As Fig. 63 illustrates, CSMA performs better than MSAP at light traffic loads, while MSAP performs better at higher loads. Because the difference in performance of the two protocols at light loads will not be great, MSAP is a better overall choice than CSMA.

It is clear from the comparisons made thus far that the appropriate choice of an access protocol is very much dependent on the environment in which it is used. For example, under light traffic conditions with a large number of users, the preceding conclusion is reversed. Hence, the authors provide a series of graphs for comparing the various protocols. The graphs display the best protocol (lowest time delay) at a given throughput for different values of N and a (a is the normalized propagation delay). The first graph (Fig. 64) displays this trade-off for the light traffic case ($S = 0.3$).

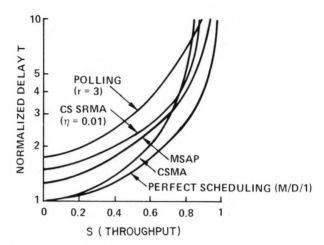

Fig. 63. Time delay versus throughput ($N = 50$, $a = 0.01$). (Based on Kleinrock and Scholl [KLEI77b].)

We note that four regions are delimited in which each of the protocols indicated provides the lowest delay. For a large population of users, random access techniques provide the best performance. As the value of a increases, the performance of CSMA degrades below that of slotted ALOHA. The MSAP is the preferred choice for a small- to medium-size population that is not widely dispersed

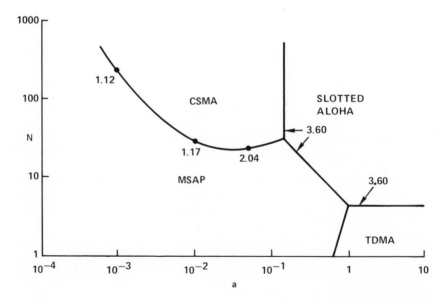

Fig. 64. N versus a ($S = 0.3$, $T_{M/D/1} = 1.21$). (Based on Kleinrock and Scholl [KLEI77b].)

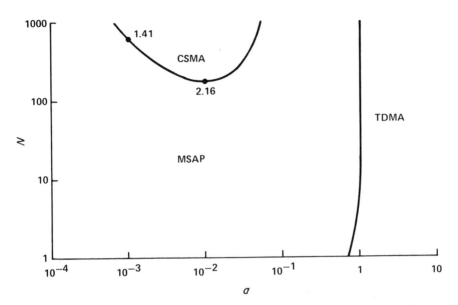

Fig. 65. *N* versus *a* (*S* = 0.6, $T_{M/D/1}$ = 1.75). (Based on Kleinrock and Scholl [KLEI77b].)

(a < 0.6), while TDMA provides the best performance for a small population that is widely dispersed.

Figure 65 illustrates the heavy traffic case of S = 0.6. Note that at heavier traffic loads, MSAP is to be preferred over CSMA even for a large number of users. When the users become widely dispersed (a ≈ 1, i.e., propagation time approximately equal to transmission time), TDMA is the preferred choice. The numbers on the boundary lines of the figures are the ratio of the time delay of the protocol to that of perfect scheduling, that is, an $M/D/1$ queue.

THE COLUMBIA* PROTOCOL

Hansen and Schwartz [HANS79] describe an assigned slot, listen-before-transmission protocol. The protocol can be made to act like nonpersistent CSMA in low-throughput regions, and to act like MSAP for higher throughput levels.

As pictured in Fig. 66, time is divided into frames, each consisting of L "minislots." The minislots are τ sec long. Assuming N users of the system, each of the minislots is assigned to a particular group of M = N/L users. This assignment establishes the priority of access to the channel.

*Both Schwartz and Hansen are at Columbia University, hence the name of the protocol.

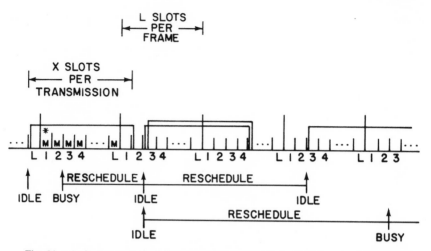

Fig. 66. Assigned-slot listen-before-transmission protocol; *, M users assigned per slot. (Based on Hansen and Schwartz [HANS79, p. 846].)

Each group of users senses the carrier on the line in the minislot determined by the assigned priority of the group. Then each member of the group follows Tobagi's nonpersistent CSMA protocol, that is,

1. If the channel is idle during the assigned slot, the user transmits his packet at the beginning of the next slot.
2. If the channel is busy, reschedule to another frame in accordance with a delay determined by a distribution. Then repeat step 1.

If all users are assigned to each frame, then the protocol is identical to nonpersistent CSMA. If one user is assigned per frame, then the system is very nearly MSAP. The difference between the MSAP and Hansen's protocol with $M = 1$ is that in Hansen's protocol a user, on finding the channel busy, does not necessarily get a turn during the next frame. Instead, he must wait a number of frames dictated by a delay distribution. The authors' results indicate that there is little difference between the performance of MSAP and the performance of their new protocol for the case $M = 1$.

Analytical and simulation models of the delay versus throughput behavior of the protocol were developed. These models will be described in the next section, while a stability analysis of the protocol will be presented in the following section.

Delay versus Throughput Models

The authors model their protocol as a discrete-time Markov process. Each user is viewed as being in one of two states—thinking or backlogged. He is in the thinking state if he has no messages to transmit, and he is backlogged if he has a message awaiting transmission.

A thinking user generates a new packet in a slot with probability σ. If the channel is busy at this time, the user enters the backlogged state and is blocked from generating new packets. He is then reassigned to each assigned slot with probability ν.

The authors define their Markov chain by considering the sequence of slots assigned to one of the user groups and, taking N_l as the random variable, the number of backlogged users in that slot.

Two separate cases are considered:

1. The number of slots X required to transmit a packet is less than L ($X < L$).
2. $X \geq L$.

The reason for separating the model into the two cases is illustrated in Fig. 67. In the event that $X < L$, then the number of backlogged users in slot l of the sequence depends on the number of backlogged users in slot $l - 1$. If $X \geq L$, then there is always the possibility that a member of the group started a transmission in a slot preceding $l - 1$, in which case the number of backlogged users in slot l depend on whether or not this transmission was successful. This possibility, illustrated in Fig. 67, precludes considering our sequence of slots as a Markov chain.

Following the authors, we first consider the case of $X < L$. In this situation, the backlog defines an aperiodic, irreducible Markov chain [PARZ62]. From a fundamental theorem on Markov processes, we have the existence of a steady-state solution to the equation

$$F\pi = \pi,$$

where F is the transition matrix of the Markov chain, and π the probability vector.

The terms in F depend on the probability of a slot being idle. In order to simplify the calculations, the authors assume that this probability is independent of

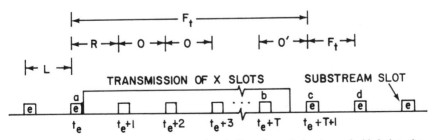

Fig. 67. Channel analysis when packet length $X > L$; e, designates embedded slots, i.e., substream slots not overlapped by a transmission by someone assigned to that substream; R, O, O', F_t, and F_l are transition matrices. (Based on Hansen and Schwartz [HANS79, p. 847].)

the backlog in the slot. (This is tantamount to assuming that traffic in different slots is independent.)

Making use of the fact that $F_{ij} = 0$ for $i \leq j - 2$, the authors solve recursively for π in terms of π_0, obtaining

$$\pi_1 = \frac{(1 - F_{00})}{F_{11}} \pi_0$$

and

$$\pi_{i+1} = \frac{1}{F_{i+1, \ i+1}} \left[(1 - F_{ii})\pi_i - \sum_{j=0}^{i-1} F_{ij} \pi_j \right]$$

for $i = 1, \ldots, M - 1$.

In the second case considered, X is assumed to be $\geq L$. To obviate the difficulties of this case, the authors take for their substream those slots NOT overlapped by a transmission from a preceding slot belonging to the same group. The transition matrix depends on the probability that an embedded slot is idle, $P(I/e)$. Making the same independence assumption as they did in the case $X < L$, the authors derive an expression for $P(I/e)$, and make use of it to calculate the terms in the transition matrix.

With expressions for π is hand, Little's theorem is used to calculate the expected packet delay D via the equation

$$D = \frac{B}{S_{out}} \quad \text{packets.}$$

where B is the expected system backlog and S_{out} the expected throughput. To calculate S_{out}, the authors note that in equilibrium,

$$S_{out} = S_{in} = E[(M - n)\sigma X] = (M - \bar{n})\sigma X,$$

where $\bar{n}$ is the average backlog for a group.

Expressions for both $\bar{n}$ and B are derived by the authors in order to evaluate D. As both the arguments employed as well as the resulting formulas are rather detailed, the interested reader is referred to Hansen [HANS78] or to Hansen and Schwartz [HANS79].

Simulation studies were also done on the protocol. When compared to the analytical model, the studies revealed agreement in the case $X < L$, but were pessimistic in comparison to the case $X \geq L$. Further simulation results indicated that the assumption of $P(I/e)$ being independent of the backlog was at fault.

In the course of performing the simulation studies, the authors studied all possible combinations of $N = (20, 50, 100, 200)$ and $X = (10, 100)$. They also compared the results to the pure and slotted ALOHA protocols, as well as to Tobagi's optimal p-CSMA protocol.

They concluded that their protocol was superior to the foregoing protocols, *if M can be chosen to optimize delay at each throughput level.* For example, at low throughputs, one would choose $M = N$ users per slot.

A comparison of the authors' protocol with pure and slotted ALOHA as well as optimum p-CSMA is presented in Fig. 68. It should be noted that the p-CSMA curve was obtained by assuming Poisson arrivals and no blocking.

Stability

The authors make use of a finite-population model in studying the stability of their protocol, basing their approach on work done by Carleial and Hellman [CARL75].

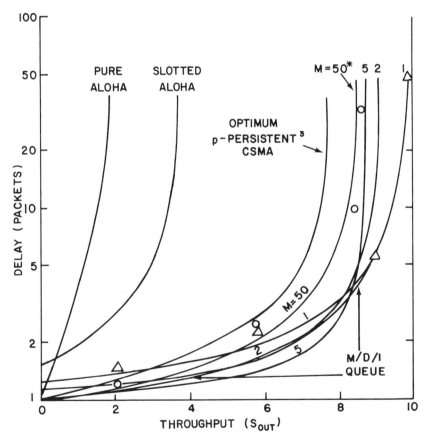

Fig. 68. Delay versus throughput $N = 50$ users, M users/slot; $X = 100$ slots/packet; $\triangle$, simulation points at $M = 1$ $(X > L)$; $\bigcirc$, simulation points at $M = 5$ $(X > L)$; *, same as slotted nonpersistent CSMA. (Based on Hansen and Schwartz [HANS79, p. 849].)

Following Carleial and Hellman, the authors define the average drift A_n to be *the expected change in the backlog of a group in a transition from a substream slot with backlog n to the next slot in the stream.* The expression for A_n is given by

$$A_n = \sum_{i=0}^{M} (i - n)F_{in},$$

where

$$F_{in} = P(N_l = i/N_{l-1} = n).$$

The F_{in} are, of course, entries in the transition matrix F.

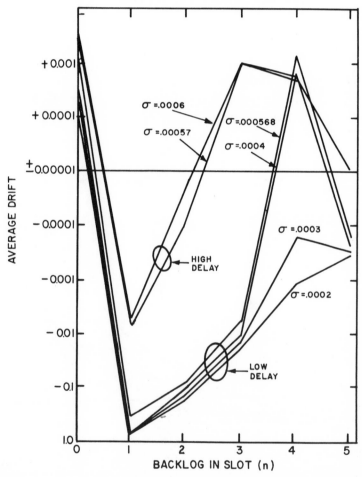

Fig. 69. Average drift versus backlog at various σ for $N = 50$ users, $M = 5$ users/slot, $X = 100$ slots/packet, $\sigma = P$(new arrival), and $\nu = .9 = P$(retry). (Based on Hansen and Schwartz [HANS79, p. 852].)

Figure 69 contains a plot of a_n versus n for various values of σ. For low values of σ ($0.0002 \leq \sigma \leq 0.0003$), the protocol has one stable equilibrium point at zero. This means that since the slope of a_n is negative, statistical fluctuations in the backlog in the neighborhood of this zero are "forced" back to zero by the negative slope of the drift. The delay versus throughput performance of the protocol is good in this region and remains so, even when the drift becomes positive. The picture changes when σ exceeds 0.0003. At some point ≤ 0.0004, two further zeros of the drift functions appear. The first is an unstable equilibrium as positive increments in the backlog around zero are still further increased, where as decrements in the backlog around zero are decreased. Once statistical fluctuations in the traffic cause the second zero to be passed, the system rapidly drifts into a region of poor performance and ultimately settles for a period of time around the third stable equilibrium point in this bad-performance region. If sufficient decrements in the traffic occur, the system eventually drifts down to the first stable equilibrium (i.e., zero).

BRAM

Chlamtac *et al.* [CHLA79] propose the BRAM (broadcast recognizing access method) protocol. As will be seen shortly, BRAM is a close relative of the Columbia protocol as well as of MSAP.

Like MSAP and the Columbia protocol, a scheduling period is employed for granting nodes access to the channel. A scheduling function assigns access to the nodes.

The scheduling and transmission periods are illustrated in Fig. 70. In this figure T_n signifies the beginning of the nth scheduling period. $H(n_1, n_2)$ is the scheduling function in which the index n_1 stands for a node which is ready to transmit, and n_2 represents the node which transmitted last.

The protocol then operates as follows:

1. If the channel is idle on being approached by a node (say j), then j evaluates $H(j, n_2)$ and schedules its transmission for time $t = T_n + H(j, n_2)a$. Otherwise, step 3 obtains.

2. If the channel is idle at time t, the node broadcasts its message. Otherwise, step 3 obtains.

3. Node j waits for the channel to become idle, and performs step 1.

The scheduling function is defined as

$$H(n_1, n_2) = \begin{cases} (n_1 - n_2 + k) \bmod k, & n_1 \neq n_2, \\ k, & n_1 = n_2, \end{cases}$$

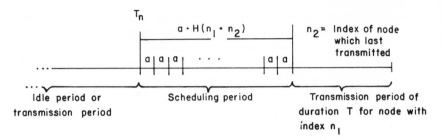

Fig. 70. Channel periods for the BRAM protocol. (Based on Chlamtac *et al.* [CHLA79, p. 1184].)

where k is the number of nodes in the network, and n_1 and n_2 are defined as previously.

Note that this function does not allow the same user to transmit successively. Hence the authors name this scheme "fair BRAM." If the preceding scheduling function is replaced by

$$H(n_1 n_2) = (n_1 - n_2 + k) \bmod k,$$

the same user can transmit twice in a row. This version of the protocol is referred to as "prioritized BRAM." Prioritized BRAM is identical to MSAP.

Both fair and prioritized BRAM suffer from the drawbacks of MSAP—at low throughput the scheduling period requires a disproportionate amount of channel time. To overcome this difficulty, the authors suggest a scheme that is identical to the Columbia protocol. The k nodes are divided into m groups and the scheduling function is defined by

$$R(n_1, n_2) = \begin{cases} (n_1 - n_2 + m) \bmod m, & n_1 \neq n_2, \\ m, & n_1 = n_2, \end{cases}$$

where n_1 is the *group* index associated with a ready node, and n_2 the *group* index of the node which was the last node to broadcast successfully.

In the event of a collision, a node waits an amount of time to rebroadcast which is determined by drawing from a uniform distribution.

The protocol bears the name of prioritized parametric BRAM.

A fair version of parametric BRAM is also defined by the authors by insisting that the first equation in the scheduling function for the prioritized version remain true for all (n_1, n_2).

Performance Models

Analytical models were developed for fair and prioritized BRAM, while simulation models were developed for all four versions of BRAM.

As prioritized BRAM and MSAP are identical, the same equations for roll call polling developed by Konheim and Meister apply. The reader will find a description of this modeling in the section on MSAP.

Fair BRAM was modeled as an $M/D/1$ queue for small values of a (<0.01). The length of the scheduling period is small compared to the transmission period when a is small; hence, the authors ignore it. Larger values of a necessitate a simulation.

The four versions of BRAM were compared via simulation with respect to their time delay versus throughput performance. The basic parameters varied were the number of nodes and a. For each throughput level, that number of nodes which minimize the delay was chosen.

The principal conclusion of these studies was that parametric prioritized BRAM offered the best performance in the medium- to high-load region, and that there was little difference between the four versions of BRAM in the low-load region. Of course, it is assumed that m can be varied by the system, something which could prove difficult under high traffic loads or under fault-present conditions.

MLMA

MLMA (multilevel multiaccess protocol) was developed at the IBM Zurich Laboratories by Rothauser and Wild. Their description of this protocol [ROTH77] forms the basis for this discussion.

The scheme is similar to Kleinrock and Scholl's AP, RR, and RO protocols in that it breaks a time frame into reservation minislots and message slots. The presumed environment for MLMA is one in which the number of users is large, but the average number of requests for channel time is small. Following the reservation frame is a message frame that consists of a variable number of packet slots, one for each reservation made in the reservation frame.

Figure 71 is a diagram of the message frame. In its simplest implementation, users request packet slots via a "one out of N" code, where N is the number of users—that is, each user sets his bit to one in order to request packet slots. As this scheme clearly becomes inefficient for large N, the authors extend their scheme to a multilevel code, which operates as follows. Let S be the base of a number system (e.g., $S = 10$); then any requestor's address R_j may be represented as

$$R_j = \sum_{j=0}^{m-1} C_j S^j.$$

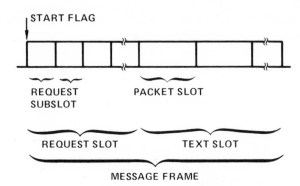

Fig. 71. Structure of the message frame. (Based on Rothhauser and Wild [ROTH77, p. 432].)

For example, $101 = 1(10)^2 + 0(10)^1 + 1(10)^0$.

The coefficients C_j (=1, 0, 1 in the preceding example) are transmitted in each request subslot. Rothauser and Wild give an example in which they claim that for $N = 1000$ terminals and $S = 10$, the length of the request slot can be shrunk from 1000 bits to 30 bits. In order to circumvent the problem of multiple requests (e.g., from terminals 013, 016, 522, and 579), the number of request subslots actually employed must be increased. Figure 72 illustrates both the problem of multiple requests and the authors' solution.

We assume requests from terminals 013, 016, 522, and 579 as before. Figure 72 illustrates the process of identifying the requesting terminals (a multilevel tree).

After the first request subslot, all users are aware that several terminals which have an address starting with either 0 or 5 are requesting space to broadcast (the exact number of requesting terminals is unknown at this point). The terminal(s)

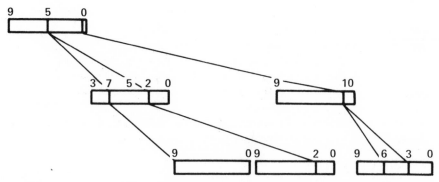

Fig. 72. Three-level tree of requestor addresses. (Based on Rothhauser and Wild [ROTH77, p. 432].)

that have inserted the highest bit (5) in the first subslot are then requested to place their second bit in the next request subslot. Bits 7 and 2 are then set as a result of this request. Repeating this procedure, the next two subslots are employed to differentiate between the last two digits of the addresses 572 and 579. As indicated on the diagram, only three slos are required to differentiate between terminals 013 and 016.

Assuming multiple requests, one may obtain a simple relationship between the number of terminals N, the number of levels in the system n, and the length of the request subslot X, as

$$N = X^n.$$

Under the same hypotheses, the channel utilization CU may be computed as

$$CU = \frac{P}{n\lceil \sqrt[n]{N} + P}.$$

where P is the packet length, $N^{1/n} = X$ is the number of bits in a request subslot, and $\lceil$ is the next largest integer. The expression in the denominator includes $n\lceil N^{1/n}$ as the overhead for each message slot. On the basis of this formula, Rothauser and Wild obtained curves such as those shown in Fig. 73 in order to determine the appropriate number of levels.

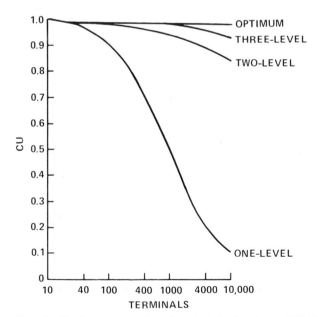

Fig. 73. Channel utilization versus number of terminals (packet size = 1000 bits). (Based on Rothhauser and Wild [ROTH77, p. 433].)

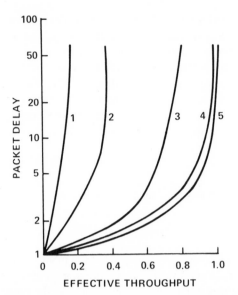

Fig. 74. Packet delay versus throughput curves: 1, pure ALOHA; 2, slotted ALOHA; 3, slotted nonpersistent CSMA ($a = 0.01$); 4, three-level MLMA ($N = 1000$, packet size = 1000); 5, perfect scheduling (M/D/1). (Based on Rothhauser and Wild [ROTH77, p. 435].)

In deriving a time-delay equation, the authors simplify the problem by neglecting (1) the size of the request slot on the grounds that it is smaller than the message slot and (2) prioritization among terminals. The result of this simplification is ideal—an $M/D/1$ queueing system, in which the average packet delay is given by

$$W = \frac{\rho}{2(1 - \rho)} \, ,$$

with ρ as the offered traffic rate.

The authors also compare their protocol with several others via the time-delay/throughput trade-off shown in Fig. 74.

GSMA

GSMA (global scheduling multiple access) was developed by Mark at the IBM Watson Laboratories [MARK78].

The feature that distinguishes GSMA from the other protocols introduced so far is the fact that it is centrally controlled. As with the systems already discussed, time frames are separated into reservation and message slots (called status and data slots by Mark). A diagram of a typical message frame is presented in Fig. 75.

The ith user is assigned a status slot s_i bits long, which he uses to indicate how many packets he has in his buffers, as well as his priority status. The data slots are used to transmit P bit-long data packets. Mark assumes that the status slot is used solely to provide information as to the number of packets awaiting transmission. Hence, each user can report a total of $2^{s_i} - 1$ packets to the central scheduler. The scheduler accepts reservation requests, and informs each user of its place in the broadcast line.

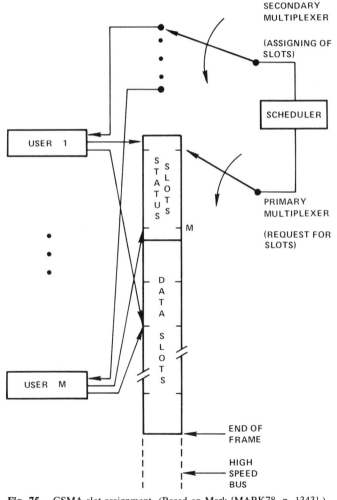

Fig. 75. GSMA slot assignment. (Based on Mark [MARK78, p. 1343].)

The system, as Mark points out, may be thought of as consisting of a single server serving each user in cyclic fashion. At the start of the jth frame, the user reports the number of packets that have arrived during the $(j - 1)$st frame in his status slots, in addition to sending the packets he requested space for in the $(j - 1)$st frame.

GSMA operates with two different formats (for the status slots)—GSMA(s) and GSMA(m). In GSMA(s), each user has a status slot of one bit, while in GSMA(m), each user has an m-bit status slot.

Mark has developed expressions for the channel utilization as well as for the time delay of GSMA(m). He assumes independent arrival processes at each node (and for each individual frame) that are governed by identical distributions of the form

$$f_i(k) = P(x_{i,j} = k),$$

where $x_{i,j}$ is the number of data units arriving at the ith user during the jth frame. Then an expression for $n_{i,j}$, the number of packets that remain to be transmitted by the ith user during the jth frame, is

$$n_{i,j} = n_{i,j-1} - q_{i,j-1} + x_{i,j}, \qquad i = 1, \ldots, M,$$

where $q_{i,j-1}$ is the number of packets transmitted during the $(j - 1)$st frame. A constraint on this equation is that

$$q_{i,j} \le n_{i,j}, \qquad i = 1, \ldots, M, \quad j = 1, \ldots.$$

This equation of the evolution of the processes is unfortunately non-Markovian in nature (the transition probability is dependent on the history of the process). Hence, using Kendall's approach [KEND53], Mark employs an embedded Markov chain to develop his queueing models.

Letting τ_j $(j = 1, \ldots, \infty)$ represent the instants in time at which the frames commence, Mark defines the cycle time t_j to be

$$t_j = \tau_j - \tau_{j-1}, \qquad j = 1, \ldots, \infty.$$

Since $n_{i,j+1}$ depends on $n_{i,j}$, it follows that $t_{i,j+1}$ depends on $t_{i,j}$. Noting that Konheim investigated a similar system [KONH76] and discovered the influence of this dependency to be negligible, Mark makes the assumption that the set of cycle times t_j $(j = 1, \ldots, \infty)$ is independent, and therefore the set (τ_j) forms an embedded Markov process.

In developing an expression for the channel utilization, the expression for the average number of customers $\bar{L}$ is used,

$$\bar{L} = \sum_{i=1}^{M} \bar{q}_i,$$

where $\bar{q}_i$ is the mean number of packets reported by the $i$the user, given by

$$\bar{q}_i = \sum_{k=0}^{2s_i-1} kPq_i^{(k)}.$$

$Pq_i^{(k)}$ is the probability that $q_i = k$.

The channel utilization for GSMA(m) is then given by

$$\mathrm{CU}_{\mathrm{GSMA}(m)} = \frac{\overline{L}P}{\displaystyle\sum_{i=1}^{M} s_i + \overline{L}P}.$$

Expressions for the ith user's mean queue size and mean time delay are also derived. (As usual, the expression for the mean time delay is a consequence of Little's theorem.) As these expressions are rather cumbersome, the reader is referred to Mark [MARK78].

Comparison of GSMA(s) time delay-throughput performance to that of synchronous time-division multiplexing (STDMA), polling, and finally to perfect scheduling yields Fig. 76. In developing these curves, Mark assumed equal arrival rates at all of the nodes. The curves for STDMA and polling are based on analytical results developed by Lam [LAM76] and Konheim and Meister [KONH74].

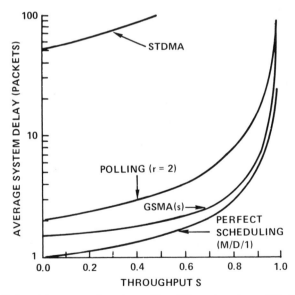

Fig. 76. Packet delay versus throughput; $M = 100$, packet size $= 100$. (Based on Mark [MARK78, p. 1349].)

DYN

Kleinrock suggests the dynamic reservation scheme (DYN) [KLEI77a]. His purpose in developing this access scheme was twofold:

1. To create a scheme capable of good performance at high traffic levels (better, e.g., than MSAP).
2. To create a scheme whose behavior does not depend heavily on how widely the users of the system are dispersed.

His description and analysis of this scheme is cursory, and is, in his own words, "basically intended to illustrate that dynamic control schemes do exist which have rather efficient behavior in the heavy traffic [KLEI77a, p. 550]." Hence, only a brief summary of the scheme and the attendant analysis will be presented.

Under this protocol, a channel of capacity C is divided into a reservation channel of capacity C_R and a data channel of capacity C_D. The capacity C of the channel is therefore given by $C = C_D + C_R$. Each user sends a brief message over the reservation channel and thereby gains control of the data channel until he is finished. Service on the data channel is first-come first-serve. The reservation channel operates in a TDMA mode. Since it is a broadcast channel, any activity on the channel indicates a desire to transmit, and a one-bit message is all that is required to reserve the data channel.

Kleinrock develops an expression for the average time delay of a packet, defined as the average time from the time the packet is generated until it is successfully received. The time delay is measured in packet transmission times—that is, one divides the actual time delay by b/C, where b is the packet length and C the channel capacity. The time delay is also defined as a function of load S on the channel, equal to $\lambda b/C$, where λ is the sum of the arrival rates at the users.

The formula developed is the sum of three terms:

1. The reservation time required, shown to be equal to $(M + 2)C/2bC_R$, where M is the number of users, b the packet length in bits, and C_R the capacity of the reservation channel.
2. The transmission delay on the data channel shown to result from an $M/D/1$ queueing system [SCHO76]. This is equal to $(2 - S_D)C/2C_D(1 - S_D)$, where S_D is the load on the data channel and C_D the capacity of the data channel.
3. The propagation delay a on the channel.

The normalized response time $T_{DYN}(S)$ is therefore given by

$$T_{DYN}(S) = \frac{(M + 2)C}{2bC_R} + \frac{(2 - S_D)C}{2C_D(1 - S_D)} + a.$$

Expressions are developed for C_R and C_D by first developing an upper bound for the load placed on the data channel (denoted by σ) and then employing the definitions $C = C_R + C_D$, $S_D = SC/C_D$, and $S = \lambda b/C$. The expression developed for σ is

$$\sigma = \frac{2\lambda M(1 - S_D)}{C_R C_D} .$$

The expressions for C_R and C_D are substituted into the foregoing expression for $T_{DYN}(S)$, resulting in

$$T_{DYN}(S) = \frac{1}{2(1 - S)} \left(\frac{\sigma b + 2M}{\sigma b} \right) \left[\left(\sigma \frac{M + 2}{2M} \right) \right.$$
$$\left. + \frac{(2 - S)\sigma b + 2MS}{\sigma b + 2MS} \right] + a.$$

The value of σ which optimizes $T_{DYN}(S)$ in the heavy traffic case is obtained by first taking the limit of the preceding expression as $S \to 1$, and then differentiating the result with respect to σ. To obtain the heavy traffic behavior of the protocol, it suffices to substitute the value obtained for $\sigma = 2M/\sqrt{b(M + 2)}$ into $T_{DYN}(S)$ and take the limit of $T_{DYN}(S)$ as $S \to 1$. This yields

$$\lim_{S \to 1} 2(1 - S)T_{DYN}(S) = \left(1 + \sqrt{\frac{M + 2}{b}} \right)^2 .$$

From this expression we note that the heavy traffic behavior of DYN is very good, as illustrated by Fig. 77. In this figure, M_{eff} is the ratio of $T_{DYN}(S)$ and $T_{M/D/1}(S)$ (perfect scheduling) and M is the number of users.

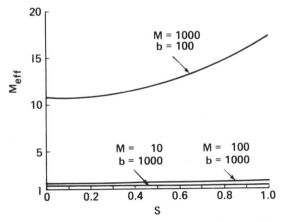

Fig. 77. Performance of the dynamic reservation scheme. (Based on Kleinrock and Scholl [KLEI77b, p. 551].)

The performance of this dynamic scheme (as illustrated by Fig. 74) is excellent unless the number of users (M) far exceeds the packet length. It also outperforms MSAP provided that a critical value for a is not exceeded [SCHO76]. DYN and MSAP, as well as other protocols, are compared in the summary section of this chapter.

THE URN SCHEME

The urn scheme is an adaptive scheme for distributed multiple access to broadcast systems. It varies from slotted ALOHA at light loads to an asymmetric scheme at intermediate loads, and finally results in time-division multiple access (TDMA) at heavy loads. It was developed at the University of California at Los Angeles by Kleinrock and Yemeni [KLEI78].

Operating under a distributed control environment, the scheme assumes that each user knows how many users are busy at the beginning of a time slot. Methods for gathering this information will be summarized later. We commence with a summary of the access scheme *assuming that this information is available to each user*.

An example of how the scheme works serves to illustrate the general solution to the problem. We assume two users ($N = 2$), and first consider the case in which one is busy ($n = 1$). If both users are given access rights (symmetric transmission policy) at the beginning of a slot, then one will broadcast and the other will not.

A symmetric policy is not always ideal, as is shown by the case of two users, both busy. Giving each equal transmission probabilities (0.5) results in a 0.5 throughput, as only two out of the four possible outcomes of this policy can produce a successful transmission. Hence, it is better to allow only one of the users to transmit with probability 1.

Yemeni [YEMI78] proves that the optimal strategy for assignment of access rights is asymmetric—that is, several users should be allowed full access rights, while the remainder should be allowed none. This is in contrast to symmetric policies, in which each user has equal access rights.

The model on which Yemeni's scheme is built is the classic probability situation of drawing colored balls from an urn (hence the name of the scheme). The traditional colors of black and white are chosen to represent the busy and idle users. The urn contains N (number of users) balls, of which n are assumed to be busy. If k balls are drawn from the urn, then the probability of a successful transmission is the probability of drawing one black ball from the urn. This is given by the hypergeometric distribution as

$$\binom{k}{1}\binom{N-k}{n-1} \Big/ \binom{N}{n}.$$

It can be shown that this probability is maximized when $k = \lfloor N/n \rfloor$, where $\lfloor x \rfloor$ is the integer part of x. This model is generalized by Yemeni [YEMI78].

Where there is only one busy user ($n = 1$), the preceding formula implies that $k = N$—that is, all the users have full access rights (although only one user will make use of them). This is essentially slotted ALOHA. As the load (n) increases, k decreases, allowing fewer users access to the channel. When $n > N/2$, only one user is allowed access ($k = 1$), and we have a TDMA scheme (random TDMA if the sampling is random, and round-robin TDMA if it is without repetition). Hence, when traffic is light, a small amount of channel capacity is lost due to collisions as the scheme operates in a slotted ALOHA mode. However, as the traffic load increases, collisions are gradually eliminated, thus maintaining a high-channel utilization.

Kleinrock and Yemeni [KLEI78] suggest a scheme for estimating the number of busy users. The scheme consists of a binary erasure reservation subchannel that may be implemented via a minislot at the beginning of each data slot. An idle user that turns busy sends a message of a few bits in this slot. Each user is assumed able to detect the presence or absence of a reservation as well as an erasure (resulting from colliding messages). Simulation studies show that assuming there are always two users involved in a collision is an effective way of estimating the number of busy users.

The specific k users that actually get access rights at any given instant can be selected by a variety of means. For example, the same random number generator that selects k out of N users may be employed by each user. The authors suggest a scheme they call a "window scheme," which reduces k when collisions occur.

Both analytic and simulation studies were conducted on the performance of the urn scheme. Its performance was compared to that of

1. optimally controlled slotted ALOHA (transmission probability $= 1/n$),
2. random TDMA, and
3. perfect scheduling.

Assumptions for both models include the following:

1. Arrival distributions are time independent, and arrivals occur independently of the slots.
2. Service mechanisms are time independent and are independent of the arrival process.

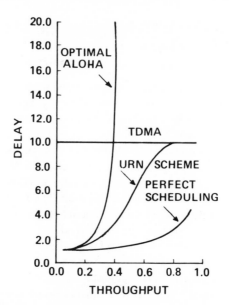

Fig. 78. Delay–throughput comparative analysis; $N = 10$. (Based on Kleinrock and Yemini [KLEI78, p. 7.2.4.].)

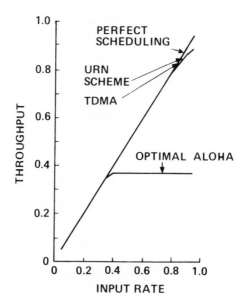

Fig. 79. Input–throughput analysis; $N = 10$. (Based on Kleinrock and Yemini [KLEI78, p. 7.2.4].)

In order to determine an expression for the time delay of the system, a model was first constructed for the number of busy users in the system. The transition equation is

$$(\pi_0, \pi_1, \ldots, \pi_N) = (\pi_0, \pi_1, \ldots, \pi_N) \begin{bmatrix} \alpha_1^0 & \alpha_2^0 & \cdot & \cdot \cdot \cdot & \cdot & \alpha_{N+1}^0 \\ \alpha_0^1 & \alpha_1^1 & \cdot & \cdot \cdot \cdot & \cdot & \alpha_N^1 \\ 0 & \alpha_0^2 & \cdot & \cdot \cdot \cdot & \cdot & \cdot \\ & 0 & \cdot & \cdot \cdot \cdot & \cdot & \cdot \\ & & 0 & & \alpha_0^N & \alpha_1^N \end{bmatrix},$$

where π_n are the equilibrium probabilities and α_i^n the transition probabilities. Solving these equations and then employing Little's theorem, the authors obtain curves for the delay-throughput performance (Fig. 78) and for the input-throughput rate performance (Fig. 79). Simulation studies have provided confirmation of these graphs.

MODELING OF BUS NETWORKS— CONCLUDING THOUGHTS

In this section, to the extent possible, the various bus network access schemes discussed in this book are compared. The performance measure employed is time delay versus throughput trade-off. In addition, the models used in evaluating the performance of the access protocols are discussed.

Comparison of Access Protocols

Any comparison of access schemes must include as a very important factor the environment in which the protocol is to be employed. Kleinrock, in an important paper [KLEI77a], mentions basic parameters that serve to characterize a distributed local broadcast communications system. They are

  1. the number of users M attached to the system;
  2. the loads S_m generated by each user;
  3. the (normalized) propagation delay a, which serves as a measure of the dispersion of the users of the system; and
  4. the total system load $S = \sum_m S_m$ placed on the entire system.

Kleinrock's system description may be simplified by noting that the last parameter is obtained from knowledge of the first two sets of parameters.

TABLE 3

The Price for Distributed Sources[a]

	Collisions	Idle Capacity	Overhead
No control (e.g., ALOHA)	Yes	No	No
Static control (e.g., FDMA)	No	Yes	No
Dynamic control (e.g., reservation systems)	No	No	Yes

[a] Based on Kleinrock [KLEI77a, p. 548].

Kleinrock goes on to note that nature exacts a price for distributed control in the form of collisions, idle capacity, or control overhead. Table 3 illustrates this price for three categories of systems.

A systems designer's problem is therefore to decide which price (or combination of prices) is appropriate to pay for his particular environment (as characterized by Kleinrock's three parameters). The performance measure apparently most appropriate for making this decision is the time delay versus throughput trade-off* made for differing values of M, a, and S. Charts similar to those presented in the discussion of MSAP (Figs. 64 and 65) should form the basis of a decision as to which protocol should be employed for a given system. Figures 64 and 65 illustrate how dependent the choice of the appropriate protocol is on the particular environment (characterized by the parameters S, M, and a.)

Kleinrock compares a number of multiaccess protocols. In doing so, he first defines a "new" system parameter M_{eff} by the formula

$$M_{eff} = \frac{T_x(S)}{T_{M/D/1}},$$

where $T_x(S)$ is the time delay of system x at load S and $T_{M/D/1}$ the time delay for the ideal deterministic system. One may view this ratio as equal to an effective number of system users via the following observations [KLEI74], referred to as the "scaling law" with C/M. The scaling law states that if we compare the time delay for the following two categories of systems:

1. M systems, each with capacity C/M and input rate S/M,
2. a single system handling total input rate S with total capacity C,

then the second system would have an average time delay equal to $1/M$ of the time delay of the first system.

A corollary of this law is that an FDMA system has a time delay that is M

---

*Stability is an important consideration for random access protocols. However, dynamic control procedures have been developed, which produce close-to-optimal, stable channel delay–throughput performance [LAM74 or KLEI76].

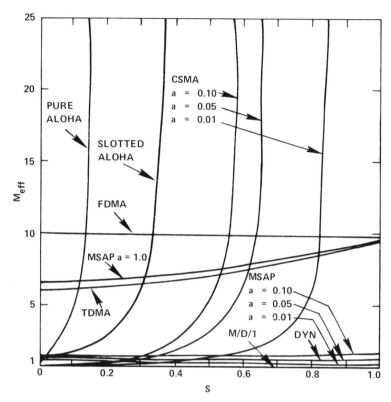

Fig. 80. Response time ratios ($M = 10$, $b = 1000$). (Based on Kleinrock [KLEI77a, p. 551].)

times as large as the deterministic $M/D/1$ system. As an example of this law, we take a system of 100 users each provided with an average 150-μsec time delay. If an $M/D/1$ system can provide each user with a 3-μsec time delay, then in effect the $M/D/1$ system provides 50 users ($=M_{eff}$) with the same 150-μsec time delay as the (proposed) system provides one user.

This new parameter* M_{eff} provides a reduced system description—M_{eff} users provided load S, and distributed at distance a.

Employing M_{eff} as a system parameter, Kleinrock presents the time delay versus throughput trade-off for a wide range of systems. Three separate cases are presented in Figs. 80–82:

1. 10 users with a packet length b of 1000 bits (Fig. 80),
2. 100 users with a packet length b of 1000 bits (Fig. 81),
3. 1000 users with a packet length b of 100 bits (Fig. 82).

*Since loads are not uniformly distributed in a system, one must allocate the system capacity in such a way as to minimize the time delay.

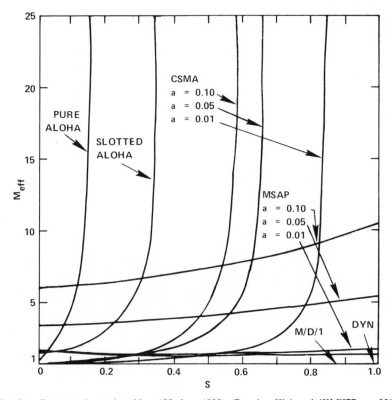

Fig. 81. Response time ratios ($M = 100$, $b = 1000$). (Based on Kleinrock [KLEI77a, p. 551].)

DYN stands for the dynamic reservation scheme discussed previously. These figures illustrate (among other things) that of the schemes compared, either MSAP or DYN is best in the heavy traffic end.

Figures 80–82 do not afford a definitive comparison of broadcast access schemes. A number of protocols are not included. Among these are

1. LWT protocol,
2. prioritized CSMA (e.g., Hyperchannel),
3. urn protocol,
4. MLMA,
5. GSAP.

It might also be desirable to portray a more detailed picture of the effect of the parameter a on these access protocols (as is done for the graphs illustrating MSAP).

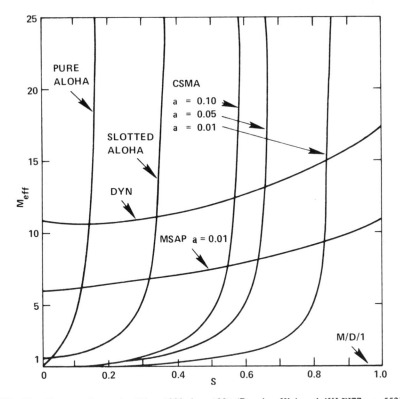

Fig. 82. Response time ratios ($M = 1000$, $b = 100$). (Based on Kleinrock [KLEI77a, p. 552].)

Performance Models of Access Protocols

Queueing and simulation models have been created for all the access protocols examined in Chapter 3. In general, there has been good agreement between both categories of models. Predictably detailed problems arose which required the use of simulation.

1. In the Hyperchannel simulation studies conducted at Lawrence Livermore Laboratories, the authors pointed out a number of problems caused by the interaction of the user-level protocols with the access-level protocols. These effects could not have been portrayed by queueing formulas that modeled only the access-level protocols. Queueing models of the higher level protocols were not developed, in all likelihood, due to the complexity of the interaction of the two protocol levels.

TABLE 4

Parameter Values[a]

Access method	Z	A	P
FDMA	2	$M/2$	1
TDMA	$(2 + M)/2$	1	1
$M/D/1$	2	$1/2$	1
MSAP	$(2 + a(M + 1))/(1 + a)$	$(a + 1)/2$	1

[a] Based on [KLEI77a, p. 548].

2. The time-delay formulas developed by Tobagi for his CSMA protocols proved difficult to optimize. Hence, he employed simulation to obtain a time delay/throughput trade-off.

In attempting to develop a unified queueing model for bus networks, Kleinrock has developed a good three-parameter approximation for a large number of the time-delay formulas [KLEI77a]. His generic formula for time delay as a function of throughput (S) is

$$T(S) = A\frac{Z - S}{P - S},$$

where Z is the zero of the function, P the pole of the function, and A a scalar multiplier. Table 4 includes the appropriate values for parameters in four access schemes. The ZAP approximation may also be employed on the ALOHA and CSMA schemes by fitting the values of the parameters.

Little work has been done on the modeling of a complete network employing a bus architecture (i.e., modeling of the nodes as well as the communications system). A potentially effective approach to the modeling of such a system would be to use Jackson's open queueing networks to model the nodes [KLEI76], and to use the appropriate time-delay formula to model the performance of the bus itself. The ZAP approximation might be effectively employed here.

In spite of the quick results to be obtained by such an approach, there is also danger resulting from a lack of detail. Hence, one might construct a two-tiered model. The first level, consisting of the preceding queueing model, would be suitable for a rapid time delay versus throughput analysis. The second level would consist of a simulation of the bus communications network, coupled to a (Jacksonian) model of the nodes as an open network of queues.

References

ABRA73   N. Abramson, The ALOHA system, in *Computer Communication Networks* (N. Abramson and F. Kuo, eds.), Prentice-Hall, Englewood Cliffs, New Jersey, 1973.

ANDE72   R. R. Anderson *et al.*, Simulated performance of a ring-switched data network, *IEEE Trans. Comm.* **COM-20,** 3 (June 1972), 576–591.

ANDE75   G. A. Anderson and E. D. Jensen, Computer interconnection structures: Taxonomy, characteristics, and examples, *Comput. Surveys* **7,** No. 4 (December 1975), 197–213.

AVI63   B. Avi-Itzhak and P. Naor, Some queueing problems with the service station subject to breakdown, *Oper. Res.* **2,** No. 3 (1963), 303–320.

BABI77   G. A. Babic *et al.*, A performance study of the distributed loop computer network (DLCN), *Proc. Comput. Networking Symp., December 1977* pp. 66–75.

BASK75   F. Basket *et al.*, Open, closed, and mixed networks of queues with different classes of customers, *J. Assoc. Comput. Mach.* **22** (April 1975), 248–260.

BEIZ78   B. Beizer, *Micro-Analysis of Computer System Performance.* Van Nostrand Reinhold, Princeton, New Jersey, 1978.

CARL75   A. B. Carleial and M. E. Hellman, Bistable behavior of ALOHA-type systems, *IEEE Trans. Comm.* **COM-23,** 4 (April 1975), 401–409.

CARS77   R. T. Carsten *et al.*, A simplified analysis of scan times in an asymmetrical Newhall loop with exhaustive service, *IEEE Trans. Comm.* **COM-25,** 9 (September 1977), 951–957.

CARS78   R. T. Carsten and M. J. Posner, Simplified statistical models of single and multiple Newhall loops, *Nat. Telecomm. Conf., 1978* pp. 44.5.1–44.5.7.

CHLA79   I. Chlamtac, W. Franta, and K. D. Levin, BRAM: The broadcast recognizing access method, *IEEE Trans. Comm.* **COM-27,** 8 (August 1979), 1183–1189.

CLAR78   D. D. Clark, K. T. Pogran, and D. P. Reed, An introduction to local area networks, *Proc. IEEE* **66,** No. 11 (November 1978), 1497–1517.

COX62   D. Cox, *Renewal Theory.* Methuen, London, 1962.

DONN78a   J. E. Donnelley and J. W. Yeh, Simulation studies of round-robin contention in a prioritized CSMA broadcast network, *Univ. of Minnesota Conf. Local Comput. Networks, 3rd, Minneapolis, Minnesota, October 1978.*

DONN78b   J. E. Donnelley and J. W. Yeh, Simulation of round-robin contention in a prioritized CSMA network, *Univ. of Minnesota Conf. Local Comput. Networks, 3rd, Minneapolis, Minnesota, October 1978.*

FARM69   W. D. Farmer and E. E. Newhall, An experimental distributed switching system to handle bursty computer traffic, *Proc. ACM Symp. Problems Optimization Data Comm. Systems, Pine Mountain, Georgia, October 1969* pp. 1–34.

FISH73   G. S. Fishman, *Concepts and Methods in Discrete Event Digital Simulation.* Wiley, New York, 1973.

FRAN77   W. R. Franta and M. B. Bilodeau, Analysis of a prioritized CSMA protocol based on staggered delays, Tech. Rep. 77-18, Computer Science Dept., Univ. of Minnesota (September 1977).

FRAN78   A. Franck, private communication (March 27, 1978).

FRAS74   A. G. Fraser, Spider—A data communications experiment, Comput. Sci. Tech. Rep. 23, Bell Labs., Murray Hill, New Jersey (1974).

FUCH70   E. Fuchs and P. E. Jackson, *Comm. Assoc. Comput. Mach.* **13,** No. 12 (1970), 752–757.

GORD80   R. L. Gordon, W. W. Farr, and P. Levine, Ringnet: A packet switched local network with decentralized control, *Comput. Networks* **3** (1980), 373–379.

HANS78   L. W. Hansen, An assigned-slot listen-before-transmission protocol for a multiaccess data channel, Eng ScD dissertation, School of Eng. and Appl. Sci., Columbia Univ. (January 1978).

HANS79   L. W. Hansen and M. Schwartz, An assigned listen-before-transmission protocol for a multiaccess data channel, *IEEE Trans. Comm.* **COM-27,** 6 (June 1979).

HAYE71   J. F. Hayes and D. N. Sherman, Traffic analysis of a ring switched data transmission system, *Bell System Tech. J.,* **50,** 9 (November 1971), 2947–2977.

HAYE74   J. F. Hayes, Performance models of an experimental computer communication network, *Bell System Tech. J.,* **53,** No. 2 (February 1974).

JACK63   J. R. Jackson, Jobshop—Like queueing systems, *Management Sci.* **10,** No. 1 (1963), 131–142.

JAFA77   H. Jafari, A new loop structure for distributed microcomputing systems, PhD dissertation, Oregon State Univ., Corvallis, Oregon, 1977.

JAFA78   H. Jafari, J. Spragins, and T. Lewis, A new modular loop architecture for distributed computer systems, in *Trends and Applications 1978: Distributed Data Processing,* pp. 72–77. Nat. Bur. Std., Gaithersburg, Maryland.

JENS78   E. D. Jensen, The Honeywell experimental distributed processor—an overview, *Computer* **11,** No. 1 (1978), 28–40.

KATZ74   S. Katz and A. G. Konheim, Priority disciplines in a loop system, *J. Assoc. Comput. Mach.* **21,** No. 2 (1974), 340–349.

KAYE72   A. R. Kaye, Analysis of a distributed control loop for data transmission, *Proc. Symp. Comput.-Comm. Networks Teletraffic, Poly Inst. Brooklyn, April 1972* pp. 47–58.

KEND53   D. G. Kendall, Stochastic processes occurring in the theory of queues and their analysis by the method of imbedded Markov chain, *Ann. Math. Statist.* **24** (1953), 338–354.

KLEI74   L. Kleinrock and F. A. Tobagi, Carrier sense multiple access for packet switched radio channels, *Proc. Internat. Conf. Comm., Minneapolis, Minnesota, June 1974* pp. 21B-1–21B-7.

KLEI75a   L. Kleinrock, *Queueing Systems, Vol. I: Theory.* Wiley (Interscience), New York, 1975.

KLEI75b   L. Kleinrock and S. Lam, Packet switching in a multiaccess broadcast channel: Performance evaluation, *IEEE Trans. Comm.* **COM-23,** 4 (April 1975), 410–421.

KLEI75c   L. Kleinrock and F. A. Tobagi, Packet switching in radio channels: Part 1—Carrier sense multiple-access modes and their throughput-delay characteristics, *IEEE Trans. Comm.* **COM-23,** 12 (December 1975).

KLEI76   L. Kleinrock, *Queueing Systems, Vol. II: Theory.* Wiley (Interscience), New York, 1976.

KLEI77a   L. Kleinrock, Performance of distributed multi-access communications systems, in *IFIP77*, pp. 547–552. North-Holland Publ., Amsterdam.

KLEI77b   L. Kleinrock and M. Scholl, Packet switching in radio channels: New conflict-free multiple access schemes for a small number of data users, *ICC Conf. Rec., Chicago, Illinois, June 1977* pp. 22.1-105–22.1-111.

KLEI78   L. Kleinrock and Y. Yemini, An optimal adaptive scheme for multiple access broadcast communication, *ICC 78* pp. 7.2.1–7.2.5.

KONH74   A. G. Konheim and B. Meister, Waiting lines and times in a system with polling, *J. Assoc. Comput. Mach.* **21,** No. 3 (July 1974), 470–490.

KONH76   A. G. Konheim, Chaining in a loop system, *IEEE Trans. Comm.* **COM-24,** 2 (February 1976), 203–209.

LABA78   C. E. LaBarre, Analytic and simulation results for CSMA contention protocols, ESD-TR-79-126, Electron. Systems Div., AFSC, Hanscom AFB, Massachusetts (May 1979), ADA070672.

LABE77   J. Labetoulle *et al.,* A homogeneous computer network, in *Computer Networks,* pp. 225–240. North-Holland Publ., Amsterdam, 1977.

LAM74   S. S. Lam, Packet switching in a multi-access broadcast channel with application to satellite communication in a computer network, PhD thesis, Rep. No. UCLA-ENG-7429, School of Eng. and Appl. Sciences, Univ. of California, Los Angeles, April 1974.

LAM76   S. S. Lam, Delay analysis of a packet-switched TDMA system, *Nat. Telecomm. Conf. Rec., Dallas, November 1976* pp. 16.3-1–16.3-6.

LIU77   M. T. Liu *et al.,* Traffic analysis of the distributed loop computer network (DLCN), *Proc. Nat. Telecomm. Conf., December 1977* pp. 31:5-1–31:5-7.

LOOM73   D. C. Loomis, Ring communication protocols, Tech. Rep. 26, Univ. of California, Irvine, January 1973.

LUCZ78   E. C. Luczak, Global bus computer communication techniques, *Proc. Comput. Networking Symp.* pp. 58–67. Nat. Bur. Std., 1978.

MAJI77   J. C. Majithia and J. D. Duke, Analysis and simulation of message-switched loop data networks, *Proc. IEE* **124,** 3 (March 1977), 193–197.

MANN78   E. G. Manning and R. W. Peebles, Distributed data processing: Case studies, *Proc. Design Distributed Data Proc. Systems, Nice, France, 1978* pp. 1–36.

MARK78   J. W. Mark, Global scheduling approach to conflict-free multiaccess via a data bus, *IEEE Trans. Comm.* **COM-26,** 9 (September 1978), 1342–1352.

MART78   J. Martin, *The Wired Society.* Prentice-Hall, Englewood Cliffs, New Jersey, 1978.

MAUR79   R. Mauriello, L. Bloom, and R. J. Malinzak, A distributed processing system, in *Trends and Applications: 1979 Distributed Data Processing,* pp. 1–10. Nat. Bur. Std., Gaithersburg, Maryland.

MEIS77   N. B. Meisner *et al.,* Dual-mode slotted TDMA digital bus, *Proc. Data Comm. Symp., 5th, September 1977* pp. 5-14–5-18.

METC76   R. M. Metcalfe and D. R. Boggs, Ethernet: Distributed packet switching for local computer networks, *Comm. ACM* **19,** No. 7 (July 1976), 395–404.

MINI80   P. Minnucan, Xerox, Intel and DEC join forces for local area network, *Mini-Micro Systems* pp. 17–20 (1980).

MUNT72   R. R. Muntz and F. Basket, Open, closed, and mixed networks of queues with different classes of customers, Tech. Rep. No. 33, Stanford Electron. Lab., Stanford Univ. (August 1972).

NESS78   D. M. Nessett, Protocols for buffer-space allocation in CSMA broadcast networks with intelligent interfaces, *Univ. of Minnesota Conf. Local Comput. Networks, 3rd, October 1978.*

PARD77   R. Pardo, M. T. Liu, and G. A. Babic, Distributed services in computer networks: Designing the distributed loop data base system (DLDBS), *Proc. Comput. Networking Symp., December 1977* pp. 60–65.

PARZ62   E. Parzen, *Stochastic Processes.* Holden-Day, San Francisco, 1962.

PENN78   B. K. Penney and A. A. Baghdadi, Survey of computer communication loop networks, Res. Rep. 78/42, Dept. of Computing and Control, Imperial College of Science and Technology, London, England.

PIER72a   J. R. Pierce, Network for block switching of data, *Bell System Tech. J.* **51,** 6 (July–August 1972).

PIER72b   J. R. Pierce, How far can data loops go? *IEEE Trans. Comm.* **COM-20** (June 1972), 527–530.

RAWS78   E. G. Rawson and R. M. Metcalfe, FIBERNET: Multimode optical fibers for local computer networks, *IEEE Trans. Comm.* **COM-26,** 7 (July 1978), 983–990.

REAM75   C. C. Reames and M. T. Liu, A loop network for simultaneous transmission of variable-length messages, *Proc. Ann. Symp. Comput. Architecture, 2nd, January 1975* pp. 7–12.

REAM76   C. C. Reames and M. T. Liu, Design and simulation of the distributed loop computer network (DLCN), *Proc. Ann. Symp. Comput. Architecture, 3rd, January 1976* pp. 124–129.

RICH79   T. G. Richardson and L. W. Yu, The effect of protocol on the response time of loop structures for data communications, *Comput. Networks* **3** (1979), 57–66.

ROBE73   L. G. Roberts, Dynamic allocation of satellite capacity through packet reservation, *AFIPS Conf. Proc., Nat. Comput. Conf. 42, June 1973* pp. 711–716.

ROTH77   E. H. Rothauser and D. Wild, MLMA—A collision-free multi-access method, in *IFIP77,* pp. 431–436. North-Holland Publ., Amsterdam.

SCHO76   M. Scholl, Multiplexing techniques for data transmission over packet switched radio systems, PhD thesis, Dept. of Comput. Sci., Univ. of California, Los Angeles, September 1976.

SCHW77   M. Schwartz, *Computer-Communication Network Design and Analysis.* Prentice-Hall, Englewood Cliffs, New Jersey, 1977.

SHER70   D. N. Sherman, Data buffer occupancy statistics for asynchronous multiplexing of data in speech, *Proc. IEEE ICC, 1970* **2,** 28-26–28-31.

STAN79   J. A. Stankovic and A. VanDam, Research directions in distributed processing, in *Research Directions in Software Technology* (P. Wegner, ed.), pp. 611–640. MIT Press, Cambridge, Massachusetts, 1979.

THOR75   J. E. Thornton *et al.,* A new approach to network storage management, *Comput. Design* (November 1975), 81–85.

TOBA74   F. A. Tobagi, Random access techniques for data transmission over packet switched radio networks, PhD thesis, Comput. Sci. Dept., School of Eng. and Appl. Sci., Univ. of California, Los Angeles, UCLA-England 7499 (December 1974).

TROP79   C. Tropper, Models of local computer networks, ESD-TR-80-111, Electron. Systems Div., AFSC, Hanscom AFB, Massachusetts (April 1980).

WOLF78   J. Wolf and M. T. Liu, A distributed double-loop computer network (DDLCN), *Proc. Texas Conf. Comput. Systems, 7th, 1978* pp. 6.19-6.34.

WOLF79a   J. J. Wolf, Design and analysis of the distributed double-loop computer network, PhD dissertation, Ohio State Univ., 1979.

WOLF79b   J. J. Wolf, B. W. Weide, and M. T. Liu, Analysis and simulation of the distributed double-loop computer networks, *Proc. Comput. Networking Symp.* pp. 82-90. Nat. Bur. Std., 1979.

YEMI78   Y. Yemeni, On channel sharing in discrete-time packet switched, multiaccess broadcast communication, PhD thesis, Dept. of Comput. Sci., School of Eng. and Appl. Sci., Univ. of California, Los Angeles, 1978.

YU76   L. W. Yu, Design and analysis of an adaptive loop-type computer communications network, MA Sci thesis, Univ. of Waterloo, 1976.

YU79   L. W. Yu and J. C. Majithia, An adaptive loop-type data network, *Comput. Networks* **3** (1979), 95-104.

YUEN72   M. L. T. Yuen *et al.*, Traffic flow in a distributed loop switching system, *Proc. Symp. Comput.-Commun. Networks Teletraffic* pp. 29-46. Polytechnic Press, New York, 1972.

ZAFI73   P. Zafiropulo, Reliability optimization of in multiloop communication networks, *IEEE Trans. Comm.* **COM-21,** 8 (August 1973), 893-907.

ZAFI74   P. Zafiropulo, Performance evaluation of reliability improvement techniques for single-loop communications systems, *IEEE Trans. Comm.* **COM-22,** 6 (June 1974).

Index